Family Money

家庭理财

(英)威廉·怀特海德◎著
(英)马克·比奇◎插图
傅瑞蓉◎译

目录 Contents

4~5 让我们来谈谈钱吧
家人之间的对话经常离不开钱，毕竟每个人都需要钱。

6~9 家庭预算
预算要先满足对必需品的需求，如果还有多余的钱，再考虑购买奢侈品。

10~15 巨大的支出
谁决定家庭的主要支出？谁付钱？

16~21 公用事业
首先要保证水、电、气的支出。

22~29 公共服务行业
你所在地区的市政当局要保证你的安全，同时还要让你能够接受教育和使用娱乐设施。

30~31 保持健康
谁支付看医生的钱？谁支付叫救护车的钱？

32~33 洗洗和擦擦
家里会产生很多垃圾。谁来负责消灭细菌？用什么消灭细菌？

34~37 餐桌上的食物
每周的食物支出是一笔大开支。超市的免费班车总是人满为患吗？

38~39 家用汽车
家里维持一辆车要付出很多成本：税收、保险、燃油、保养……

40～41　花园
　　　　花园需要细心照看，有时还可能需要专家的帮助。

42～43　宠物
　　　　宠物也要看医生，而且养宠物的成本可不低！

44～47　家庭娱乐
　　　　如果没有任何娱乐，生活又有什么趣味？家庭度假或出国旅游可能是一笔大开支。

48～49　养育你要付出的成本
　　　　养大一个孩子（你）需要一大笔钱，多得令你吃惊！

50～51　你的教育
　　　　教育可能是免费的，也可能是不免费的，但是无论如何你的父母都要付出一定的代价。

52～53　你的服装
　　　　如果你是一个赶时髦的人，你应该知道新衣服有多费钱。

54～55　你的零花钱
　　　　你的零花钱也是家庭预算的一个重要组成部分，你可能很需要它。

56～57　总结
　　　　这里提供了一份家庭预算范本，你可以照着给你家做一个预算。

58～59　讨论
　　　　钱不会从树上长出来。你家为什么会花掉那么多钱？

60～61　中英文术语对照表

62～63　索引

　　64　译后记

附　英文影印版

让我们来谈谈钱吧

你是否听你的爸爸妈妈说过很多次,想做这个,想做那个,但是可惜没有足够的钱?他们是负担不起啊!有关钱的话题以及如何花钱这个问题总是经常被我们的家人所提起,因为有了钱,才能买到我们家庭所需要的所有东西。钱也关系到你自己生活的方方面面,例如,你的零花钱也是家庭开支的一部分。

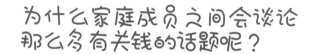

为什么家庭成员之间会谈论那么多有关钱的话题呢?

很简单,钱的多少决定了你能否舒适安逸地生活,它也决定了你将以何种方式进行生活。之所以大家会如此频繁地谈到钱,是因为无论钱够不够花,家庭中的每个成员或多或少地都需要一些钱。

家庭成员相互之间的交流,可以帮助大家明白自己可以拥有什么,不可以拥有什么,并且让大家明白,当我们可以购买某些自己想要的东西时,应该心存感激。

钱进来了

你的爸爸，或者你的妈妈，或者父母双方，会有一份工作。他们每个星期都会花上一定的时间从事一份特定的工作，他们或者在某一个办公室里，或者在某一个商店里，或者在其他一些地方，甚至有可能在家里工作。

如果他们是按每周的工作时间、按商定的费率得到报酬的，那么，这个报酬就被称为工资或薪金。在每一个周末或每一个月的月底，你的父母将会收到他们的报酬。这些报酬通常是直接打进家庭的银行账户里的。

你的父母亲是做什么工作的？

钱出去了

家庭所必需的东西，包括吃的、穿的以及取暖，都需要花钱，家庭计划要的东西也需要花钱。食物以及汽车的燃料需要花钱，电费、煤气费……所有这一切都需要花钱。如果全家人想外出吃饭，那就得付餐费，去看电影就得付电影票的钱。甚至许多家庭还得为旅行或度假而进行储蓄呢！

那么，到底应该如何正确地花钱呢？

家庭预算

你的父母是如何知道什么东西是他们可以负担得起的,什么东西是他们负担不起的?他们对钱是如何进行分配的?大多数父母都会制定一个支出模式,来决定多少钱将用于购买每周或每月的家庭必需品。把那些真正重要的需求,以及他们的花费以清单的形式列成表格,这就叫预算。

谁负责做家庭预算?

你的父母知道,如果他们在其中一个项目上花掉了太多的钱,那么,他们能够用于另外一个项目的钱就变少了,因此,密切关注家庭预算是一项非常重要的工作。

你要问一问你的爸爸或妈妈,他们两个谁是家庭财务预算"主管"。也许这个预算是他们两个人一起做的。许多爸爸妈妈都是一起做家庭预算的,因为要尽量做到双方都满意。

钱够花了吗？

或许你的家庭是没有预算的。或许家里的每个人都只知道花钱，而且希望钱永远也花不光。但是，这真的是一个好的计划吗？

这样做或许在一段时间内是没问题的，不过随后就很可能会发生意想不到的事情。家里的汽车坏了，屋顶开始漏水了，或者你的爸爸还是妈妈生病了（并且因此而不能去工作了），这时，家庭的预算就会变得非常紧张。家里的每个人，当然也包括你，都需要明白发生了什么事情。

必需品

必需品和奢侈品

奢侈品

你的家庭每月都会有一定的开支，要帮你买衣服、付取暖费，还要为你购买食物，所有这些都是必需品的支出。这些都是预算当中的关键项目。

在你的家庭收入当中，可能会有一部分多余的钱，它可以花在那些你希望自己最好能够拥有的东西上。也就是说，那些东西是你想要的，但并不是必需品，这些东西就被称为奢侈品。

没有预算，就很容易超支；超支了，就意味着你们的家庭会陷入债务当中。

债 务

你欠债了，意思就是说你欠别人钱了。你也许欠的是你的朋友或者父母亲的钱，他们或许根本不关心你什么时候还钱。但是，几乎可以肯定的是，当一个家庭欠债时，关于什么时候还钱这个问题，那个债主必定是关心的。

米考伯先生的故事

著名的英国作家查尔斯·狄更斯写了一个故事。他在故事里塑造了一个名为米考伯的人物，他花钱如流水，总是入不敷出。

后来，米考伯先生由于欠债太多而被投进了监狱。

但他吸取了自己的教训，并且给别人提出了一个很好的建议……

如果你的年收入为1英镑，如果你的年度支出为99便士，那么，这个结果就是幸福的。

但是，如果你的年收入为1英镑，你的年度支出却为1.01英镑，那么，这个结果就是悲惨的。

银行贷款

如果你的家庭陷入了暂时无法偿还的债务当中，或者想筹集一笔钱去度假或购买一件特别的商品，你的父母亲就会去跟银行谈，以求获得贷款。银行会想知道，你们是否还得起这笔贷款，并且会设定一个必须偿还的期限。银行还会收取一笔被称为"利息"的费用。一般来说，这笔费用将会被算进这笔贷款当中，而且必须定期偿还。

信用卡

借钱还有另外一种方式。但是利用这种方式与利用银行贷款比起来，你必须付出高得多的利息，即信用卡透支。这似乎是得到额外的钱的一个简单的方法，但是它仍然是一种债务，仍然是必须偿还的。

这是我的问题吗？

债务几乎总是会引起一个家庭的焦虑和困难。如果你明白这是怎么一回事，并且明白为什么会发生，那么，你就会对你的家庭做些有所帮助的事情。你可以勒紧你的裤腰带，或者你也许会变得更耐心一些、愿意多等等。

巨大的支出

你现在住的房子可能是你父母利用按揭贷款买下来的,或者也有可能是租来的。除非你的父母完全拥有这幢房子的所有权,否则住房成本将是你父母亲的一笔巨大的开支,也是他们的一个巨大的预算项目。平均来说,它会占到你的家庭收入的25%左右。

什么是按揭贷款?

按揭贷款是指你向银行或其他机构借款,以帮助你购买你所住的房子。大多数银行会让你在25年左右的时间里付清你所有的按揭贷款。这笔贷款需要你每月平均偿还一小部分。

银行要收取一定的利息。该利息将与按揭贷款的本金加在一起,需要你定期偿还。

什么是租金?

有些房子是租来的。租金是指住房者支付给实际拥有房屋的人或者房东的一笔费用。租金通常是按月支付的。

租借双方会签署一种叫做租约的租赁协议,租用期限一般为几个月到几年不等。根据租赁协议,一旦租赁期满,租方就得续签租赁协议,或者搬离这栋房子,另找住处。

保险是什么？

你的家，你家里的所有家具、厨房电器以及停在车库里的汽车，这些东西都值很多钱。当然，它们都有可能会出故障，会坏掉，甚至有可能会被偷走——任何时候都有可能。火灾、风暴以及暴雨导致的洪水，所有这些意外灾难都有可能会让你家的财产遭受损失。

你的父母可能会选择为你的家以及你家里的一切财产进行投保。这就意味着你的父母每年将支付一小笔钱给保险公司，而当你家里的什么东西需要维修或更换时，保险公司就会帮你家偿付。

维修和保养费

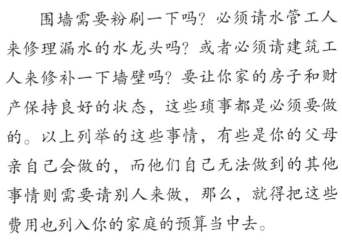

围墙需要粉刷一下吗？必须请水管工人来修理漏水的水龙头吗？或者必须请建筑工人来修补一下墙壁吗？要让你家的房子和财产保持良好的状态，这些琐事都是必须要做的。以上列举的这些事情，有些是你的父母亲自己会做的，而他们自己无法做到的其他事情则需要请别人来做，那么，就得把这些费用也列入你的家庭的预算当中去。

养老金

年老体弱

总有一天，你的父母会停止工作，一旦这一天到来，他们的固定工资或薪金支付就将停止发放了。然而，他们很明智，他们在工作的时候会从每个月的工资里拿出一小部分钱存起来，他们会持续这样做好多年。

这笔钱是每个月从他们的工资或薪金里扣出来，然后交给政府代为保管的。或者，你的父母亲也可能把他们用来养老的钱交给其他养老保险公司代为保管。

那么，什么是养老金？

上述那些储蓄被称为养老金。你的父母亲退休之后，那笔钱就将会返还给他们。养老金是一种投资，是会钱生钱的，因此养老金的总额每年都将会稳步增加。最终合计的总数将会远远多于从他们当初工资里扣去的钱的总额。

利息的作用

如果你把一笔钱存在银行里,或者把一笔钱借给别人做生意,那么,你会希望这笔钱多起来,这就是说,你希望当这笔钱还回来的时候,会多出一点点来。

确保在还款时钱会多出来的一种方法是,收取一定的利息。利息是支付给借款人的利润或奖励;如果没有利息,借款人就不会有动力去投资,钱也不会多起来。

计算利息有很多种方法,但是最简单的一种方法叫做单利计算法,它是在一个基准金额的基础上按照固定的利息率计算的。

年份	投资额	利率为每年 10%	总计
1	1 英镑		1.10 英镑
2			1.20 英镑
3			1.30 英镑
4			1.40 英镑
5			1.50 英镑

税 收

你肯定听到过你父母亲对缴纳所得税的抱怨。这是因为它是法律规定的，如果你不纳税，你将会受到重罚。但是，即使人们会抱怨，但他们还是知道为什么必须缴纳所得税。所得税是指每个人从自己的工资中拿出一定比例的钱缴纳给政府。通常情况下，有钱的人缴纳得多一些，穷人则缴纳得少一点，有些穷人甚至不用缴税。但是，几乎所有的人都必须拿出一部分钱来存入政府的"储蓄罐"里。

销售税

政府也可能会对商店里销售的商品征收一定的税。你可能已经注意到了，当你买东西的时候，它的价钱会比标签上注明的价格高一些。这些销售税可能是缴纳给联邦政府的，也可能是缴纳给州政府的。销售税可以高达商品价格的20%。

这些税款将花在哪些地方？

大家之所以愿意纳税，是因为税款取之于民，用之于民。政府把税款用于国民所需要的各种服务上，比如说医疗保健、国防、教育和养老金等。

政府的钱的去向

并非所有的政府都是以同样的方式花钱的，但是大多数政府一般都会花在以下这些方面。

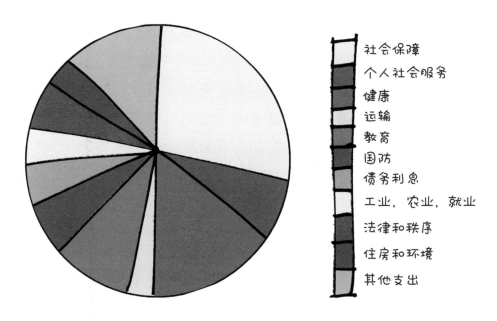

社会保障
个人社会服务
健康
运输
教育
国防
债务利息
工业，农业，就业
法律和秩序
住房和环境
其他支出

公用事业

公用事业有哪些？

"公用事业"是一个非常有用的术语，它是指政府提供一些基本的服务，例如被输送到千家万户的电、天然气和水。它也指某些公司提供一些基本的服务。

由谁提供这些基本服务呢？

提供这些服务的是公用事业公司，它们确保电灯能够亮起来，电视能够打得开，它们也要确保我们的火炉能够燃起来，水壶里的水能够煮沸，或者我们的水龙头里能够流出干净的水来。

在有些国家，这些基本服务是被政府部门控制的，或者是由政府部门提供的。它们对每个人的健康和安全都非常重要，因此，政府非常关注这些基本服务的提供情况以及它们的费用支出。

享受这些基本服务要花多少钱？

提供这些基本服务的公司是不可能免费为你供电、供热和供水的，它们是要向你收费的。

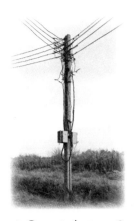

这是一个电网，通过架设电缆把电力从一个城镇输送到另一个城镇

电力从巨大的发电站输送出来，然后穿越千山万水送达你家

生活污水在污水处理厂被处理干净，干净到足可以重新进入供水系统

从深海处钻出石油，并且通过输油管道把它输送上岸

你什么时候付费？为什么要付费？

在你家的某个地方安装有一个家庭仪表箱，它是为记录你家所使用的每一项公用服务而安装的。这个仪表箱能够算出你家到底使用了多少服务，它也能够显示你家必须支付多少费用。每隔几个月，便会有人来读取这个仪表上的数据（或者由计算机自动读取）。这些读数显示了你家用了多少服务，以及你该支付多少费用。仪表的数据一旦被读取出来，不久之后账单也就送到你家了。

如果你不交费会怎么样呢？

供水、供气和供电的公司都希望你能够按时交费。如果你稍微延迟一点交费，它们是会等等的，但是最后，它们会对你发出警告。如果你再不交费，它们将停止供应这些服务。它们真的会这么做！

公用事业：取暖和照明

电力

电是一种能源，它能够让你的电灯发光，让你厨房里的电器运转起来，它还能让你打开你的电视。电的测量单位是瓦。电是由当地发电站通过电线输送到你家的。

石油

石油所"走"过的道路是很漫长的。它先从油田里被钻探出来。油田一般都位于地表深处或者深海处。然后输油管道或巨大的超级油轮再把它从一个国家运送到另一个国家。有时候它要穿越半个地球呢！

电能可以通过燃烧石油、天然气和煤炭而产生。我们也可以通过核能发电，或者通过让有巨大落差的水流驱动涡轮来发电。电力通过架设在空中的电线，穿越田野和山区，进行远距离的传输，最后送达你所在的地区。在乡镇和城市里，电既可以通过地下，也可以通过架设电线杆在空中传输到你家。

水

你有没有想过水是从哪儿来的？你冲洗马桶的水或者洗碗后的水又都到哪里去了？水是由自来水公司供应的，有时候还是由你所在地的小镇供应的。

你家里用的洁净的饮用水是从管道里流出来的，而这些管道是与埋在地底下的巨大管道连接在一起的。这个巨大的管道被称为总管道。总管道里的水来自于几千米（甚至更远）之外的蓄水池。这个蓄水池可能是一个湖泊，也可能是一个被称为含水层的地下水库。

污水从水槽、浴室和厕所里排出去后，流进了你家地下室的一个大管道里，这个大管道穿行于地底下，并且连接到了埋在街道下面的一种被称为下水道的更大的管道里。

污水最后流向了污水处理厂，污水处理厂在去除了这些污水的臭味、灭杀了细菌后，又重新把它排放回了地面，或者让它流进河流和大海。

天然气

许多家庭是使用天然气来取暖和做饭的。它要么从巨大的气缸里通过管道输送到你家，要么通过卡车把它送到你家的储气罐里。跟石油一样，它也是从地底下被开采出来的，并且也是经过远距离的运输才能送达你家的。

公用事业：让你我保持联系

今天，大多数人都希望能够随时随地与自己的家人、朋友通话，也希望能够随时随地与自己的工作场所、商店和供应商保持联系，还希望无论自己在哪里，不管是在什么时间，都能够进行娱乐休闲活动。如今，全球化的公司既为我们提供了这种机器，也为我们提供了这种连接服务——当然这得付费。

电话

电话是从装在墙上的接线槽里连接到你家的。墙上（或墙内）装有双绞线，双绞线穿过房子一直连到了外面的电线杆上。这条线会经过许多根电线杆，有时候它还会穿行于地底下，最后与中心交换站连接到一起。通过这个中心交换站，它把信号发送到全球各地的交换站。

移动电话

移动电话的信号来自于发射塔，发射塔遍布城乡各地，每隔几千米就会安装一个。如果语音信号需要传播得更远些，那么，这些信号就会被传送到绕着地球转的卫星上，然后通过卫星再传送到全球各地的发射塔网络上。今天的移动电话几乎可以做计算机能做的所有事情。

互联网

互联网是一个全球性的计算机网络系统。计算机用户能够与全世界各地的亿万其他用户进行交流。万维网使用的就是互联网的结构，它能够利用浏览器访问被链接上网络的文档。网页包含有文字、图片、声音和视频，用户可以通过点击链接，对它们进行交互浏览。

电视

输送电视信号的电缆线与互联网的电缆线是一样的。会有那么一天，电视与计算机会合二为一。

信件和包裹

许多国家的邮政系统可以追溯到数百年前，邮递员每天都会送信件。但是现在我们一般都用电话和电脑进行沟通，利用信件进行沟通的人已经越来越少了。

邮递员现在已经很少了。

公共服务行业

在每个村庄、小镇和城市,都会有那么一些人,他们为我们的某些需要提供服务。这些事情是我们自己无法完成的。比如说搬运垃圾、办学校和图书馆、灭火等。这些活动都被统称为公共服务,提供这种服务的人通常都在公共建筑内工作。

公共服务项目主要有以下这些

* 清除垃圾与街道清洁
* 消防部门
* 警察
* 道路
* 环境
* 公共卫生
* 公园
* 图书馆
* 幼儿园
* 老人
* 残疾人
* 博物馆
* 活动中心

地方税

提供这些公共服务都是需要花钱的,而这些钱主要来自于各种各样的地方税。这些地方税有时候与你所在地区的房子的价值有关(即房产税),或者也可能取决于生活在这个地区的人数的多少(即人头税)。这些税款被用于学校的建设、聘请教师以及提供学校教育;它还被用于购买消防车,以备不时之需;它还会被用于购买搬运垃圾的垃圾车,以及支付做这些工作的工人的工资。

垃圾清运

我们每天都会把废纸和其他垃圾扔进家里的垃圾桶里。厨房里也会制造出许多垃圾，比如说一些瓶瓶罐罐和食品包装物，它们也会被扔进垃圾桶。所有这些垃圾都会被分门别类地装进不同颜色的袋子里，这些袋子有纸做的、塑料做的，也有用金属和玻璃做的，所有这些被分装好的垃圾都会被放入垃圾桶里。仅就一个家庭的垃圾，都有可能堆成大大的一堆。

每周一次，垃圾都会消失不见。垃圾车会定时过来把垃圾运走。在几个小时内，这些垃圾要么被回收再利用，要么被运到巨大的垃圾填埋场，然后在那儿让它们自个儿腐烂。

垃圾在腐烂发臭或者危及人的健康之前，必须被处理掉

成千上万吨垃圾被倾倒在垃圾填埋场

公共服务行业：道路

谁在照管我们的街道和道路？每个地区都有一个部门叫做公共工程部门。这个部门专门负责填洼补坑、清扫污垢和修补沥青道路，还负责在寒冷的天气里用铲雪车消除路面积雪，以疏通道路。

在任何一个城市，干净的街道意味着健康的环境

在空中，到处都满布着各种电线、电话线、网络线和电视线，它们都附着在电线杆上。各种管道、排水沟则横躺在地下。街道是社区的生命线，它承载着人们所需要的各种服务。公共工程部门确保公用事业和其他组织能够正常地运作。

电线和电话线并不总是整洁干净的，但是它们对我们的家庭和工作来说，都是必不可少的

地方公共卫生部门

还有另一个部门来专门负责当地的卫生健康工作。他们的工作是确保人们能够呼吸到清洁的空气、喝上干净的水，以及使用安全的建筑材料来建造住房、厂房和道路。这个部门还专门负责检查餐馆厨房里的卫生状况，同时监测空气污染情况。

市政当局把钱用于什么地方？

钱必须被用于建设……

幼儿园和学校

从很小的时候开始，你就会与你家附近的其他孩子一起接受教育。

博物馆

你可以在当地的博物馆和美术馆里看到你最喜爱的动物、恐龙，了解各国历史知识和欣赏艺术。

关爱残疾人

残疾人在学校里、工作场所和家里都会有一些特殊的需求，他们需要市政当局的帮助。

公园

公园是一个开放的场所，那里绿树成荫，还有许多运动游乐场所，大家可以在那儿尽情地玩耍。

关爱老人

老人并非全都是能够生活自理的，他们有些需要养老院的帮助。

体育运动中心

参加体育运动中心的某一团队或者进行锻炼。

图书馆

在图书馆里有成千上万的书籍和其他媒体读物供借阅，那里同时还有舒适的座椅。

公共服务行业：消防部门

我们大家都着迷于消防车。每当消防警报响起时，都会让人激动不已，大家都会自动地让出道路让它通过。

当房子、工厂、办公楼和商店着火时，消防车便会出动，它带着长长的云梯，能够伸到高高的屋顶上。它还带着高压水龙头，能够迅速扑灭大火。这些高压水龙头会被接到街道上的消防栓里，而消防栓是连着总水管的。我们的饮用水也是从这个总水管里流出来的。

当发生森林大火、道路车祸时，或者有人在家里摔倒了，现场也会出现消防车或消防救护车的身影，它们甚至还会营救被困在屋顶上的宠物猫。

消防工作

如果你想成为一名消防员,那么首先要进行一些艰苦的训练。你还必须足够聪明,已经通过了正常的学校教育。但是无论如何,你必须身强体壮。试着想象一下,在一幢被大火燃烧着的大楼里,你要沿着楼梯把一名昏迷不醒的伤者安全地从三楼救出时的情形。

用来灭火的水来自最近的消防栓

训练

作为一名消防员,你要接受很多训练。你必须学会如何扑灭各种不同类型的火。有些火用水就能扑灭,而有些火则需要一些特殊的喷雾剂才能扑灭。消防员必须学会使用各种特殊的设备,比如消防水带、云梯、灭火器和一些消防工具。

火灾的危害

由火灾引起的一个最致命的问题是烟。消防队员必须学会:需要穿戴什么样的护身设备,如何爬行通过狭小的空间,如何扑灭高层建筑的大火,如何处理危险的材料和化学品,等等。

高架云梯使消防员能够到达高层建筑的高处去灭火

伸出援助之手

在紧急情况下,可能还需要消防员进行现场医疗急救;即使不需要,消防员也必须保持冷静,以协助现场的营救工作和予以其他方面的帮助。

各种消防车辆都将到达火灾现场

公共服务行业：警察

无论你走到哪儿，你都会发现有警察在那里维持秩序。他们会保护你，使你不会遭小偷偷窃，不会被拦路抢劫者抢劫。他们也会保护你不受持有危险武器的不法分子的侵害。事实上，任何一个想伤害他人或者破坏他人财产的人，都会受到警察的追捕。

警察要做很多工作。有时候，如果你停车时间过长或停错了地方，交通警察就会给你开罚单。如果你以危险的方式驾车，那么，交通警察就会阻止你，或者要你缴纳罚款。

当然，警察也会在其他方面帮助你。他们会帮助你过马路，或者建议你应该往哪儿走。他们都是受过训练的，工作时他们会像朋友一样对待你和保护你。

在法庭上

如果一个人被指控为罪犯，那么，他必须让法庭做出判决。如果发现他真的有罪，那么，他就可能会被罚款或被投进监狱。法庭这样做是要付出一定成本的，它所花的钱也来自于各种不同的税款，其中包括你父母所缴纳的税款。

警察工作

警察的日常任务是帮助他人,以及时刻保持警惕。

无论是在马背上,还是在快速摩托车上,警察都在观察着周围的一切。

个人技能

如果你想当警察,你将需要掌握许多不同的技能。而且要想成为一名合格的警察,还要通过许多考试。警察的工作很紧张,因此要想成为一名警察,需要很大的勇气,而且你得一直保持身体健康,并始终充满力量。

警察需要知道所有这些知识

警察必须掌握基本的医疗知识,以应付突发事件;警察还需要掌握一定的法律知识,以便在法庭上为他们自己的行为做出解释;他们还必须知道如何富有同情心,能够与各种各样的人打交道。

自卫

警察需要学会如何使用枪支;要懂得一些防卫措施,包括搏击术;还要学会如何应对危险的情况和危险的人,要知道如何不让自己和他人陷入危险的境地。

保持健康

让家里每个人都保持健康是父母亲最关心的问题之一。除了不明原因的咳嗽和感冒外，几乎所有儿童通常容易犯的疾病，父母亲都要确保能够用正确的方式进行护理，而这种护理的费用有时候是很昂贵的。

家庭医生

大多数家庭都会有家庭医生，他们通常是当地社区的医生或者一些实习医生。在家庭医生那里会有你的一份医疗档案，这份档案详细记录了从你出生开始一直到现在的你的所有病史。医疗档案的内容包括了对你的所有治疗情况、用药情况和住院治疗情况等。

当地的牙医那里也会有一个相同类型的文件，在这个文件里列出了所有有关你的牙齿的治疗情况。这些记录意味着，当你生病时，医生是在许多已知事实的基础上决定如何给你治疗的。

但是，所有的这些记录、医生的拜访以及所使用的药品都是需要花钱的，这些钱或许是你的父母以某种特别的纳税方式支付的。

治病付钱

在许多国家,政府有责任确保人们保持身体健康。它要求每个人都从自己的工资里拿出一小部分钱来支付保持身体健康所需的花费。这种服务有时候被称为国民卫生保健。

药品

治疗

如果你参与了国民卫生保健计划,那么,无论你何时得病,你都可以去看医生。如果你的病情很严重,或者如果你发生了意外,那么,你可以得到免费的住院治疗,甚至可以免费叫救护车。

牙齿

通常,儿童的牙齿护理与其他医疗保健一样,都是免费的。你在年幼时好好护理你的牙齿是很重要的,你的牙医会告诉你应该如何护理。

当你生病时,你可能需要一些药品来帮助你恢复健康。几乎所有这些药品都是很便宜的。如果你参加了国民卫生保健计划,那么,这些药品还可能是免费的。

洗洗和擦擦

我们的家里经常会搞得一团糟。在厨房里,食物会洒落一地,油污会四溅,甚至会杯盘狼藉;浴室里会落满灰尘,浴缸和洗脸盆周围会积有污垢和头发。那么,你的卧室呢?它是不是每天都保持干净、整洁和气味清新?

清洁费

几乎每个家庭都需要每星期进行一次大扫除,这就意味着必须要有人去做。或许你的妈妈或爸爸做大扫除工作,而你也会帮着做点清扫工作。或者,也许你的父母会雇用别人来打扫卫生。如果是这样,那么,清洁费就应该纳入家庭开支的预算中。

清洁工具

每一项清洁工作都需要特殊的清洁工具:洁厕灵用来清洗卫生间和灭杀细菌;油污净用来去除油渍;空气清新剂用来让空气变得清新;洗衣粉(洗衣液)用来清洗脏衣服;而香皂和洗发水则能够让你变得干干净净。

小工具和机器

家里的清洁设备是一大笔开支。你想想，你家里有多少清洁用具。通常而言，清洁用具包括家庭清洁用的、熨衣服用的、清洗地板和墙面的，甚至可能还有其他的清洁小工具。

● ● ● ● ● ● ● ● ● ● ● ● ● ● ●

保险

家电出问题，也许是它里面的主要零件发生故障了，也许是其中的一个小零件出问题了。如果你的父母已经为它买了保险，那么，保险公司就会支付它的维修费，甚至还会为你们家支付更换小零件的费用。

保险是指为了防止发生某些不好的事情，预先定期地支付一小部分钱给一个叫做保险公司的组织。以后如果真的发生了不好的事情时，保险公司就会负责帮你支付费用。

餐桌上的食物

在一个家庭中,最大的开销之一是每星期采购食品的费用。在采购过程中,一些最基本的食品是必需的,当然你的父母还可能会购买一些"奢侈的"食品。这些"奢侈的"食品是你很喜欢的,但不是必需的,比如说冰淇淋和蛋糕。

自己种植

很久以前,农民们都是自己种植农作物的。今天,你也可以在你家的花园里种植水果、蔬菜,甚至还可以种植一些粮食作物。你可以饲养鸡、山羊,甚至还可以养一头奶牛。你自己种植的水果、蔬菜比去商店里购买便宜多了。

露天市场上的水果既新鲜又便宜

但是,你仍然需要把一些成本考虑进去。你首先需要购买种子、小动物。你需要一些工具和动物饲料。不过即使如此,你自己生产食物还是比去超市购买要便宜。

商店

虽然现在仍然还有露天市场,在那里有农民们自己种植的农产品销售,但是大多数人都会到街角小店、专门的食品店以及大型超市里去购买食物。

品牌的成本

你的父母可能会对摆放在超市里的食品的价格货比三家,他们必须做出决定,到底应该购买哪个品牌的产品。品牌是商品的特定名称或标志,不同品牌的商品价格是不一样的。如果你认出了某个品牌,那通常是这个品牌做过全国性的广告,一般来说,它的价格也会更高。商店自己的品牌会比较便宜,那是因为它不必支付在电视上、广播里和杂志上做广告的费用。

刚从花园里挖出来的新鲜的胡萝卜富含维生素

数量多少?

商品价格的多少,还要看它的容量,2升装的牛奶就比1升装的牛奶要更实惠。但是,如果大包装的食品在保质期内吃不完,那么,它就有可能变质,因此这个时候它就不一定便宜了。

是否新鲜?

几乎所有的食品都标有一个日期,它表明这个食品的新鲜程度,它会建议你最佳购买时机,并且告诉你在到期日之前把它吃完。

外出就餐

如果妈妈说她懒得做饭,而爸爸也不想做,那么,全家人可能就会到外面去就餐了。外出就餐是最昂贵的一种饮食方式,虽然在一些国家,在饭店、大排档、熟食店和快餐店吃一顿饭的价格并不高。大部分家庭都会有预算,每个月可以外出就餐一两次。

看看其中的价差

在家里制作的汉堡主要由以下几样东西组成:番茄酱、生菜、西红柿、小圆面包、奶酪。这几样东西做成一个大汉堡,它的价格为1.03英镑。

快餐店的汉堡包里面的东西和家里的一样,但价格为3.30英镑。

有升有降

因此,如何花钱,全在于我们的选择,不是吗?

不幸的是,吃、住的花费是必不可少的,关于它们,我们没有太多的选择。事实上,大多数人都发现,他们把大部分钱都花在了这些生活必需品上了,因此算出在必需品上的花费是一件非常重要的事情。

价格会上升或下降,人们的收入也会有升有降。那么,应该怎么办呢?人们非常关心一定数量的钱在各个不同的时间里到底能买到多少东西,一种方法是通过"生活成本"来衡量。"生活成本",顾名思义,就是我们的生活需要花费多少钱。

在购买食品时看看价格总是有好处的

一篮子食物

对于食品来说，衡量方法是，比较一下标准的一篮子食物的成本，看看它有什么变化。这一篮子食物的成本，每年、每个月甚至每个星期都在变化，因此做好预算是一件相当棘手的事情。

即使是加工食品，它的价格也可能有所变动，虽然生产商会在价格低廉时买到原料，但是他们不会在价格低廉时一次性把原料都买齐，通常是在很久之后，当有需要时才会去购买原料。

一切皆因天气

新鲜食品的价格，比如谷物、咖啡（一般称为农产品），通常取决于天气。就拿小麦来说，如果种植小麦的地区干旱缺水，那么就有可能歉收，因此小麦的供应量就会减少，价格也会因此而上升。依此类推，小麦价格的上涨会带动面包和谷类食品的价格上扬。

甚至连肉类的价格都会受到影响。这是因为用于饲料的粮食减少了，牛和农场里饲养的其他动物就都会被过早地宰杀，那么，到第二年，肉牛就会变少了，因此，牛肉的价格也就会上涨，同时伴随而来的是牛排和汉堡的价格也将上涨。所有这一切皆因天气的变化。

农产品可能在远未成熟时……

……且价格上涨前就被买下了

有些农产品被晾晒和风干，然后被储存起来

遭遇严重的旱灾，农作物颗粒无收

家用汽车

你可能经常会要求你的爸爸或妈妈用汽车载你到某个地方去。如果他们告诉你,他们不能载你去时,你可能会生气。明明家里有汽车,为什么我还要乘公共汽车?答案是,自己开车是一项非常昂贵的花销,燃料只是汽车费用中的一部分而已。

购车时的花费

购车最贵的一项初始费用是购置费。大多数人无法一次性付清购买一辆新车的费用,甚至是一辆二手车也不行。这就意味着他们不得不借钱购车。由于大部分钱都是借来的,因此接下来的一段时间里,他们必须分期偿还,即每个月定期地支付一部分钱。

汽车牌照

你要开车,就必须要有驾驶执照。考驾照得花费一定的费用。驾照的有效期从2年到10年不等,一旦到期,你就必须进行更换。

注册登记

每辆汽车都必须到一个专门的办公室去注册登记,以证明汽车的所有权。在一些国家,注册登记的费用要比考驾照的费用高。

税

除了燃油税之外,国家或地方政府还会把消费税加进汽车的价格中。这些税收是用来支付道路、桥梁以及高速公路的维护保养的。

燃料费

现在大多数汽车都是通过汽油或柴油来驱动的。燃料的价格取决于石油的价格和政府对燃料征收的税款。在有些国家，税收是燃料价格的四倍，因此，小型汽车或者混合动力车更省油或节省能源，这种汽车比大排量的汽车更经济实惠。

保养费

即使是新车也需要进行正确的保养，这就意味着会产生保养费。更换机油、滤清器、刹车零件和轮胎，都是要付钱的。在一些国家，若干年后，汽车必须通过安全检查，任何有磨损并且可能会出现危险的零件都必须进行更换，而更换零件费以及安全检查费都必须编入家庭开支的预算之内。

报废的汽车堆积如山，它们被当作废金属回收再利用

污染

汽车是我们这个星球上最严重的污染源之一。它们所使用的燃料是不可再生资源，成本高昂。它们释放出来的气体会损害我们的健康。它们最终会报废、会毁坏或者自然老旧，进而变成一堆堆废铁。你为什么不使用公共交通工具或采用步行的方式呢？

花 园

如果你家有一个花园，那么，在花园里种点吃的东西倒不失为一种很好的节约开支的方法——还会对地球环境有好处呢！因为这样，你将有助于减少矿物燃料的使用，也将有助于减少从世界各地运送新鲜农产品到你所在地的超市所造成的污染——飞机运送和冷藏货车都会对环境造成污染。

你只需要几平方米的土地、一些水和一点点时间就可以了。

窗台上

即使你没有一个大花园或者任何类似的园子，你仍然可以种植瓜果蔬菜等植物。如果你家有一个阳光充足的阳台或露台，或者在窗台上有一个室内药草园，那么，请考虑一下容器花园吧！当一个小小的盆子里种植出了那么多的西红柿和辣椒时，你会惊奇万分的。

改善你的健康状况

保持健康需要做的最重要的事情之一,就是多吃新鲜蔬菜和水果。自己种植的蔬菜和水果的品相更好,也更美味。蔬菜和水果的维生素含量在刚从花园里采摘下来时是最高的,你可以直接食用。

节省生活开支

如果你和你的家人食用从自己家花园里种植出来的果蔬食物,那么,你家的食品开支将会减少很多。你只需花很少的一点钱购买种子,它便能够生产出数公斤的果蔬产品。当然,你也可以自己从成熟的瓜果蔬菜中留出种子,你把这些种子晒干,到第二年就可以拿来种植了。

享受更美味的食物

新鲜食物是最好的食物。你知道超市货架上的食物已经存放在那儿多久了吗?你知道食物从农场到餐桌上要走多远的路吗?

自己种植的番茄的味道与从商店里买来的番茄的味道相比,好比苹果与墙纸的味道。很多人都非常关注食品市场销售的食品的安全问题。当然,如果你吃的是你自己种植的食物,那么,毫无疑问,你自然相信你的食物是安全的、健康的、适宜食用的。

宠　物

几乎可以肯定地说,你家将会养一只宠物。一半多的美国家庭都拥有一只以上的宠物,而在欧洲,拥有宠物的家庭超过了 7 000 万家。

当然,也许现在你的宠物很小,像一只沙鼠那么小——如果是那样的话,那就不需要花费你很多钱。但是,如果你的宠物很大,比如说是一只大猎犬或者是一匹马,那么,养宠物的成本将是一大笔开支。

欧洲人的宠物

在欧洲,有将近 2 500 万的宠物是养在家里的。在本书的第 43 页列出了英国的各种宠物数量,你能与你自己国家的宠物数量比较一下吗?

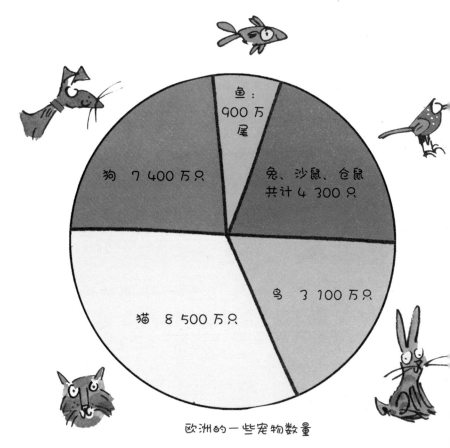

欧洲的一些宠物数量

宠物费用明细表

饲养宠物除了初始成本之外——有些宠物在购买时就花了一大笔钱——还会产生其他一些费用。以下是一张宠物费用表，这是一个普通的宠物主人在一年内花在宠物身上的钱。

据估计，一只狗的花费包括：
玩具／礼物／款待。
美容。
兽医费／医疗费。
狗舍／猫舍。
食物。
总计　1 183英镑。

一匹马的花费每年可高达4 000英镑

宠物统计

据估计，48%或1 300万个英国家庭至少拥有一只宠物。在英国，总共大约有6 700万只宠物。

狗：800万只。
猫：800万只。
养在室内的鱼：有2 000～2 500万尾鱼养在鱼缸里。
养在户外的鱼：有2 000～2 500万尾鱼养在池塘里。

兔子：100万只。
养在笼子里的鸟：100万只。
豚鼠：100万只。
仓鼠：超过50万只。
家禽类：超过50万只。
蜥蜴：30万只。
青蛙和蟾蜍：20万只。
蝾螈／火蜥蜴：10万只。

蛇：20万条。
乌龟／海龟：20万只。
沙鼠：10万只。
马／小马驹：10万匹。
鸽子：10万只。
昆虫类：不到10万条。
小老鼠：不到10万只。

家庭娱乐

平时,你的父母忙于工作,你也要在学校里努力学习。回到家里,你的父母有许多家务活要做,而你也要完成你的家庭作业。唉,总是有那么多忙不完的活要干,因此,大家的闲暇时间就变得非常宝贵了。

当然,你可能想整天泡在电脑前,或者戴上耳机沉浸于自己的世界当中,但是大多数家庭都会设法让全家人一起共度美好时光。与家人在一起的时光实在是太温馨了,根本不容错过。全家人可以一起前往冒险乐园,可以去看电影,可以去参加露天音乐会,可以一起外出就餐,也可以一起去赛马或参加足球比赛,等等。有这么多有趣的事情可以做,你还想错过吗?

体育运动是让你度过你的空闲时间的一种非常好的方法

免费娱乐

现在,很多娱乐都是免费的,但是也有许多是要花钱的。在很多情况下,你能享受多少乐趣,取决于你有多少假期以及你父母有多少空闲时间,当然还取决于你们家除去日常开支之外,还有多少余钱。

不需要花太多钱的娱乐

一个家庭要想享受美好的休闲时光,不一定需要花费很多钱。你可以在自家后花园里玩游戏,比如说踢足球;你也可以邀请一些朋友加入,一起到公园里玩游戏;你还可以骑着自行车去乡村远足或者与大家一起骑自行车穿越整个城市的街道。

闲暇时间可以与你的朋友和家人一起度过

或者你自己一个人独自度假也很好

廉价的休闲娱乐

这儿有一些建议,你可以不必花费大笔金钱,但是却可以很好地享受你的闲暇时间。

你可以举办一个聚会,邀请你所有的朋友前来,但请你的朋友自带食物和饮料。

你可以到海滩或公园里待上一整天,在咖啡馆或酒吧用午餐。

去看一场足球赛或听一场音乐会,你可以选择高处的廉价座位。虽然坐得比较靠后,但是有大屏幕,它不会让你错过任何的场景。

假日时光

假日时光的意思是，全家人可以在一起休闲娱乐。度假的时间可以选择在冬天，也可以选择在夏天或春天，通常会持续几个星期。你可以去度假胜地，也可以去海滨别墅、山间小屋，甚至还可以在游轮上度假。你度假时间的长短、度假地方的舒适程度，取决于你的家庭经济状况。

在主题公园的家庭度假，一定是一次千载难逢的美好经历

如果你想去国外旅游，那么费用就会上升。大多数家庭都会提前做好准备，这样他们就可以为这次度假编制预算，可以每个月预留出一部分钱出来。你同样也可以这么做，你可以从你的零花钱中或者你自己赚来的钱中留出一部分，建立你自己的假期基金。

廉价的旅行度假……

当许多旅游公司宣传廉价旅行度假和廉价机票时,它们实际上针对的是你预算之内的度假和机票。它们也假设你的预算是有限的。它们的真正意思是,只要你做好预算,旅行度假其实是不贵的。

廉价的旅行度假往往意味着你可以留在自己的国家,也许你可以租一辆家用小轿车,然后到处转转。也可能意味着你可以去国外旅游,但参加包价游。在包价游中,你去的大部分地方都已经为你定好了,包括机票、住宿和膳食。

……或者豪华游

或者你可以度过一个奢侈的假期。几乎可以肯定地说,你会到一些热门的度假胜地,那里的酒店和设施都很豪华。那里有海滩,酒店里有运动和娱乐设施。

总之,你如何度过你的假日时光,完全取决于你的父母亲愿意为这次假期花费多少钱,也意味着你们必须做好一个预算以确保有足够的钱可用……

……一旦你做了一个很好的预算,那么就可以保证全家有一个美好的假期。

养育你要付出的成本

你的父母肯定不会把花在你身上的钱看做是成本。他们爱你，他们会尽自己最大的努力给你幸福。他们希望你身体健康和强壮，他们想为你提供最好的学习机会，让你充分发挥自己的潜能。他们可能根本不会去计算花在你身上的钱是多少，但是现在还是让我们来算一算吧！

让我们想象一下，你的父母将要照顾你21年，从出生那天起到你21岁——21岁刚好是你大学毕业的时间。

他们不会惯坏你，但是他们会为你支付所有的基本费用或必需品费用，比如食物、衣服、校服和日常出行需要的费用。他们还会为你支付一些奢侈品的费用，诸如你参加课余的体育运动和俱乐部的费用，以及购买电脑、手机的费用。他们还会给你零花钱和生日礼物。

在英国，一对父母亲花在一个孩子身上的钱可能高达22.25万英镑。而在美国，这一数额将超过30万美元。

花在你身上的成本

教育费——校服，课后俱乐部和大学费用，交通费。

育儿费和保姆费。

食品。

服装。

度假费用。

发展业余爱好的费用和购买玩具的费用。

闲暇时的费用。

零用钱。

购买家具的费用。

个人费用。

其他。

总计　9 000 美元。

对于父母来说，养育小孩得付出巨大的成本

开发你的创造性技能是教育的一个重要组成部分

然而，一半孩子……

全世界上几乎有一半人——30多亿人——每天的生活费不足2.50美元。这些人当然也包括孩子在内。

如果一天只用2.50美元，那么在欧洲各国，养育一个小孩，一年需要大约900美元，把一个小孩从出生养育到17岁只需要1.65万美元。而全世界有一半的孩子生活在贫困当中。

你的教育

事情似乎就是这样永无休止地持续下去：起床—上学—回家—做作业，每天周而复始。有时候你真的希望这种生活有结束的一天。但幸运的是，你的父母已经跟你解释了为什么你需要过这样的生活，你的早期教育是你一生当中最重要的事情。

免费教育

对家长来说，在大多数国家早期教育通常都是免费的，这包括小学和初中，甚至是整个中等教育。建设学校和聘用教师的钱来自于你父母缴纳给地方政府以及联邦政府的税。你的父母所缴纳的税也会被用于其他方面的支出，如公共交通、教科书和体育活动。

教育成本

你的父母知道，如果你没有受过教育，那么等你长大后，就业机会就会变得非常有限，你将赚不到多少钱。所以，他们现在花在你的教育上的钱是值得的。以后，你就不要再问"我必须做家庭作业吗？"这个问题了，因为你已经知道答案了！

课后补习

在韩国，教育还包括额外的补习。全国95%的中小学生在放学后都会去参加课外补习，提供这种补习的课后辅导机构被称为"补习学校"。这些机构为孩子们进入高等学校提供帮助。也有韩国的学生去非常专业的学校进行学习，比如说武术学校和音乐学校。因此，许多学生不但学习时间很长，而且补习时间也很长，时常要学习到深夜才能回家。

大学专科和大学

你该为你的大学专科或者大学教育支付多少学费，往往取决于你的居住地。在大多数国家里，你接受大学专科或大学教育的费用，只有一小部分是由国家税收负担的，而大部分学费都是要家长自己支付的。当然，学生也可以申请助学贷款，但是到时候必须及时偿还。在英国，学生自己需要承担的大学费用是有限的，但是仍然需要数千英镑。在美国，大学教育的费用则是英国的四倍之多。而在印度尼西亚，却很便宜，它是美国的十分之一。

如果你是远离家乡去读书，那么教育成本就要增加了，你还要支付你的食宿费用。

降低高等教育费用的一个方法是获得奖学金。奖学金能为你的高等教育的学费和生活费提供额外的资金。这个奖学金可能是由政府提供的，也可能是由捐助大学的私人提供的。

你的服装

你衣柜里的衣服少得可怜吗？或者你衣柜里的衣服堆积如山？又或者你所拥有的衣服介于两者之间？你可能希望自己能够拥有一些最时髦、最昂贵的运动鞋。其实关于你的服装，唯一真正重要的事情是，够穿了就好！

校服

在许多国家，学生穿校服的时间占到了大多数。对于学生们来说，这是削减服装开支的一种方法。校服很少有外表时尚的，但是可以保证的是，它不会引起同学们之间的相互攀比，而学生家长们也不用面对孩子购买各式各样服装的压力了。校服是一种伟大的"平衡器"。

时尚品

大家都知道，你并不需要用最时髦的名牌服装来装扮自己，你不需要用服饰来衬托和提升自己。但是这也并不意味着你非得把自己弄得衣衫褴褛不可。其实你不必付出高价，就可以让你自己的穿着变得得体大方。

广告商和时尚杂志可能会建议你付高价买名牌服装，因为这是他们的工作。你必须考虑你自己或者你的父母是否负担得起。在你的家庭预算当中，你父母或许打算把家庭收入的10%拿来支付全家的服装费用——这已经是一大笔数目了。

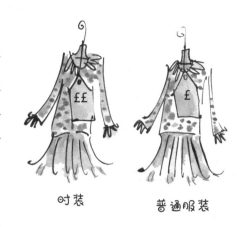

时装　　普通服装

越来越便宜

服装并不一定都是昂贵的，有些专卖店会打折销售，有些连锁店也会卖一些平价商品。

在今天，普通服装一般来说都不贵，因为现在全世界有很多服装生产大国，比如说中国和印度，它们的劳动力价格比较低廉。在这些国家，缝制一件T恤衫的成本只有美国的一半。

到廉价商店或者到能够用最少的钱买到最多的物品的地方购物

你的零花钱

一般来说，你的父母亲是不会向你收取你在家里吃住的费用的，他们还有可能会帮你购买你的学习用品。不过大多数父母还是宁愿给孩子零花钱，孩子自己个人的开支让孩子自己负责。

许多父母认为，在孩子年纪还小的时候就学会管理自己的钱是非常重要的。他们可能会定期给你零花钱，而不是当你向他们要时才给你钱。

你能得到多少零花钱取决于你的父母亲愿意给你多少钱，或者他们认为你应该需要多少钱。

零花钱的金额和获得零花钱的时间

零花钱是你获得固定收入的第一步。你可能希望你的零花钱是一个星期给一次，但这取决于你的家庭预算安排，同时取决于你这笔零花钱能用多久。如果你认为你在拿到零花钱的第一天就会把钱都花光，那么，一个星期给一次并不是一个好注意。

你的父母给你零花钱可能是有附带条件的，他们可能会要求你做一些家务。这时，你需要确切地知道你父母希望你做什么。

你的"收入"

这里所说的家务活，可能只是你的父母让你简单地在厨房里帮一下忙，或者打扫一下自己的房间。实际上这些事本就是你应该做的。他们也可能是让你照料一下宠物或者做一点花园里的工作，或者是让你在他们干活的时候帮一下忙。

没问题，你会得到报酬的！

总　结

我们把上面说的这一切都总结一下。

什么是预算？

预算是对你家里的钱是怎么来的、怎么花掉的一个估计。你家里的钱主要来自于你父母亲的工资。如果你有哥哥或姐姐，那么也包括他们赚来的钱。你家里的钱就是你的家庭收入。

在同一段时间里，花掉的钱就称为开支。在这里，我们举一个家庭预算的例子。

比方说，你的家人的工资总额是每月 4 500 英镑，如右图所示，你家里预算中的家庭开支为 3 945 英镑。

那么，剩下的 555 英镑可以作为储蓄存起来，也可以用于购买一些"奢侈品"。你们可以全家人一起去看一场电影，也可以购买全家人都想买的一些东西，或者甚至你的父母还可能会给你一点额外的零花钱。

每月收入

家庭工资 4 500 英镑

每月支出

家庭支出
 食品 450 英镑
 电费 200 英镑
 燃气费 150 英镑
 水费 35 英镑
 修理费 100 英镑
 维护费 175 英镑
 保险费 180 英镑
 共计 1 290 英镑

按揭或租金 1 500 英镑

旅游
 汽车燃料 105 英镑
 汽车维护费 150 英镑
 保险 125 英镑
 汽车贷款 275 英镑
 共计 655 英镑

税
 水 50 英镑
 地方税 450 英镑
 共计 500 英镑

总计 3 945 英镑
余额 555 英镑

下载

你不必像上面那张图所示的那样，事无巨细地把什么都写在纸上。如果你的父母并没有像左图所示的那样做预算，那么，你可以帮他们设计一个预算模板。

有一些非常容易使用的电子数据表，你可以去以下网站下载，网址是：http://www.budgetworksheets.org/.

现在，你可以请你的父母亲帮你把所有的项目明细都填上去了，然后你就可以把它们都计算出来。

讨 论

小账单

在家庭预算中,有很多方法可以帮助你减少开支。仔细想想在通常的某一个日子里,你的家庭的所有花费吧——电费、燃气费、水费、供热费、交通费、餐饮购物支出、购买衣服的支出等。对于减少家庭开支,你能做些什么呢?这里提出一些建议供参考:随手关掉不需要用的电灯;出门多走路,少乘公交车。请讨论一下你能够做到的其他一些减少开支的事情吧!

钱是长在树上的吗?

几乎每个人都希望钱是长在树上的。如果真是那样的话,那么,生活就会简单得多,我们的地球上也就会有更多的森林。然而,实际是,钱是有价值的,它可以用来交换同样有价值的商品和服务,钱是一种宝贵的东西。你花的每一分钱都必须是赚来的,这通常是你父母亲每小时辛苦工作的所得。

你怎么赚钱?

如果你够幸运,你会得到一些零花钱,你可以随意地花你的零花钱。但是许多父母都认为,孩子的零花钱必须是孩子自己赚来的。你可以通过帮助他们做些家务或者干一些跑跑腿的事情赚零花钱。那么,你做什么事情能够帮到你的家庭呢?你可以问问你的邻居或者你的朋友,他们有什么愿意付费的活让你干。

购物有问题吗?

如果你真的需要购买什么东西,那就放心大胆地去买吧,这根本没什么,但千万不要买着玩!请记住:在家庭预算里有必需品和奢侈品之分,所以你需要理智地花钱。最好的办法是列一个清单,然后坚持按清单购物。请你想想还有没有其他办法可以控制你的支出。

为什么你需要银行？

对你的父母亲来说，拥有一个银行账户是必须的，因为有了银行账户，你的父母亲才能把钱存放在安全的地方，而且还可以支付账单。你可能有一个储钱罐，你把零花钱和其他多余的钱都放在了那儿。但是你也可以把你的钱存在银行里。如果你每个星期或每个月都存一小笔钱到你的储蓄账户里，那么，银行会给你利息吗？讨论一下你如何才能开立一个银行账户，你愿意用你存的钱去买什么东西。

为什么不创建一个心愿单？在上面列明每件商品的价格，这样你就知道你需要在你的账户里存够多少钱了。你甚至可以给自己设定一个期限，规定在某一个日期之前赚到并存够你所需要的钱。

怎样保证安全和改善环境？

所有东西都要花钱，而且可能会花很多钱，所以好好地珍惜钱真的非常重要。你还记得你应该如何保护你自己、你的钱、你家的财产以及你生活的环境吗？作为初学者，你千万不要忘记，万一出了什么问题，不要慌张，还是有一些方法可以帮你弥补损失的。

此外，环境也需要我们好好保护。我们要降低生活成本，要让我们的生活环境变得更健康。请讨论一下，你有什么方法可以保护和改善你周围的环境。以下给出的是一些建议：对所有物品，你要尽可能地做到回收再利用；你可以自己种植蔬菜。你还能想到其

他方法来保护你和你的家人吗？

信用卡透支还是贷款？

你的父母亲可能会去银行办一种特殊的卡，这样他们就可以不需用现金来支付了。这类卡被称为信用卡或借记卡。你知道它们之间的区别吗？讨论一下，这两者哪种更好用，为什么？最后，如果某样东西你家里有需要，而全额付款又承担不起，那么，这时候银行会贷款给你们——比如买汽车的时候。你还记得你们家还需要其他贷款吗？

中英文术语对照表

allowance
请参见 pocket money（零花钱）。

appliances 家电
是一些做家务用的电器设备，比如水壶、洗衣机等。

aquifer 含水层
一种地下水库。

bank account 银行账户
个人与银行签的一种协议，让银行帮你管理钱。

bill 账单
供应商签发给顾客要求他支付款项的一种声明。

brand 品牌
制造商给产品定的标志或名称，很容易被人们识别出来。

budget 预算
计划花费的金额。

commodities 商品
主要指一些比较重要的产品，如谷物和金属等世界上进行大量买卖的产品。

cost of living 生活成本
基本家庭开支。不同时期的生活成本可以相互比较。

credit 信贷
借来的钱。

credit card 信用卡
能够让你用借来的钱购买东西的卡。

debt 债务
需要偿还的借款。

disabled 残疾人
身体某些部位的行动受到限制并需要特殊帮助的人。

essentials 基本生活用品
家庭的基本开支，也被称为生活必需品。

expiry date 到期日
印在食品包装上的日期，提示购买者应该在这个日期之前把食品吃掉。

fossil fuel 矿物燃料
在地球上能够找到的燃料，如煤和汽油等。

garbage 垃圾
废弃的家庭用品，能够被收集、销毁或回收再利用。

government 政府
由民众选举而产生的、代表民众管理国家的一群人组成的团体机构。

hagwon 补习学校
韩国的私人补习学校。

health care 卫生保健，医疗保健
是政府实施的一个计划，确保所有人都能享受某些医疗保健。

income 收入
也称 earnings。

income tax 所得税
是税收的一种，这种税由政府根据人

们的收入的多少来征收。

interest 利息
由于出借或投资而增加的钱，是按本金的一定的百分比计算的。

insurance 保险
有些人或公司交给保险公司的一小部分金额，当出现财产损坏或者损失时就会得到保险公司的大笔赔偿。

investment 投资
把钱借给个人或公司，帮助他们发展业务。

lease 租约
是一项协议，协议中规定在支付一定租金的情况下，允许使用一段时间。

licence 许可证，执照
用钱换来的一种许可权限。

luxuries 奢侈品
想购买但却不需要的东西。

meter 仪表
是一种小工具，用来计算家里的电费和水费的。

mortgage 抵押贷款，按揭
给房屋购买者的贷款，可以分期偿还。

pension 养老金
人们在年轻时分期投资于政府和公司的资金，等到年老的时候会得到返还。

pocket money 零花钱
家长给孩子并让孩子用于个人消费的钱。

pollution 污染
由人类的废弃物所导致的对环境的破坏。

poverty 贫穷
没有足够的钱、无法满足基本需求的一种状态。

price 价格
购买东西的成本。

public works 公共工程
由政府代表社区承担的建设活动。

recycling 回收
利用废弃物制成新的产品。

rent 租金
由于占用了所有者的财产而定期缴付的款项。

salary 薪水
因工作而赚得的固定收入。

services 服务
由地方政府出钱承办的一些项目，比如修建学校和博物馆等。

sewage 污水
由家里产生的废弃的水。

tax 税收
收入或支付的一部分，由政府征收，并用于一些公用事业。

utilities 公用事业
电力、水和其他基本生活需要的产品供应的总称。

wage 工资
因工作而得的固定收入。

61

索 引

advertising 广告 35
afford（verb）给予，提供，买得起（动词）4, 6, 38
allowance 零花钱 4, 54, 55, 56
ambulance 救护车 25, 31
appliances 电器 11, 18, 33
aquifer 含水层 19
art gallery 美术馆，画廊 25
bank account 银行账户 5
bargain 便宜货 35
bill 账单 5, 9, 17, 41
brand 品牌 35
budget 预算 6, 7, 10, 11, 32, 36, 37, 39, 42, 46, 47, 53, 55, 56, 57
cable 电缆 17, 18
car 汽车 5, 7, 38, 39
chore 家庭杂务，日常零星事务 11, 44, 55
cleaning 清洗 32, 33
clothes 衣服 32, 33, 48, 49, 52, 53
commodities 商品 37
cost of living 生活成本 36
credit 信贷 9
credit card 信用卡 9
debt 债务 7, 8, 9
defence 防御 15, 29
dentist 牙医 30, 31
diesel 柴油 39
disabled 残废的，有缺陷的 22, 25
doctor 医生 30, 31
earnings 收入 36, 55

education 教育 15, 27, 48, 49, 50, 51
electricity 电力 5, 16, 17, 18, 24, 57
essentials 生活必需品 7
expenses 支出 7, 43, 44, 52, 54, 56, 57
expiry date 到期日 35
fire 火 11, 22, 26, 27
fire hydrant 消防栓 26
food 食品 5, 23, 32, 34, 35, 36, 37, 40, 41, 43, 48, 49, 51
fossil fuel 矿物燃料 40
fuel 燃油，燃料 5, 38, 39, 40, 57
garbage 垃圾 22, 23
garden 花园 34, 40, 41, 45, 55
gas 气（煤气，天然气）5, 16, 17, 18, 19, 27, 57
government 政府 12, 14, 15, 16, 31, 39, 50, 51
hagwon 补习学校 51
health/health care 医疗保健，卫生保健 15, 16, 22, 23, 24, 30, 31, 41, 48
holiday 假期 5, 9, 44, 46, 47, 49
hospital 医院 30, 31
housekeeping 家庭开支 32
income 收入 8, 14, 53, 55, 56
income tax 所得税 14
interest 利息 9, 10, 12, 13
insurance 保险 11, 33, 57
internet 互联网 21
investment 投资 12, 13, 34
job 工作（可数名词，侧重具体职业）5,

6，11，22，28，32，44，50，53
    landfill 垃圾填埋场 23
    landlord 房东 10
    law court 法庭 28
    lease 租赁，租约 10，38
    leisure 休闲 45，49
    library 图书馆 22，25
    licence 许可证，执照 38
    loan 贷款 9，10，51，57
    local tax 地方税 22
    luxury/ies 奢侈／奢侈品 5，7，34
    mail 信件 21
    mains（water）总管道（水）19，26
    market 市场 35，41
    medecine 药品 30，31
    meter 仪表 17
    mobile phone 移动电话，手机 20
    mortgage 按揭，抵押贷款 10
    museum 博物馆 25
    nuclear energy 核能 18
    park 公园 25
    pension 养老金 12，13，15
    pet 宠物 42，43，55
    petrol 汽油 39
    pocket money 零花钱 46，48，49，54，55
    police 警察 22，28，29
    pollution 污染 24,40
    poverty 贫困，贫穷 49
    power line 电线 18
    power station 发电站 17，18
    price 价格 35，36，53
    processed food 加工食品 37
    public works 公共工程 24
    recycling 回收 23
    rent 租 10
    repairs 修理 11，33，57
    reservoir 水库 19

    restaurant 餐厅，饭店 36，44
    salary 薪水 5，12，14
    scholarship 奖学金 5
    school 学校 22，25，27，44，48，49，50，51，52，54
    school uniform 校服 48，49
    services 服务 15，16，17，20，22，24，26，28
    sewage 污水 17，19
    shop 商店 5，14，20，26，34，35
    simple interest 单利 13
    sport 体育运动 22，25，40，45，47，48，50
    supermarket 超市 35，40
    switching station 交换站 20
    tax 税收 14，15，22，28，30，39，50，51，57
    telephone 电话 20，21，24
    television 电视 21
    traffic police 交警 28
    turbine 涡轮机 18
    tutoring 辅导 51
    university 大学 48，49
    utilities 公用事业 16，17，24
    vet 兽医 43
    wage 工资 5，12
    water 水 16，17，18，19，26，27，40
    work 工作（不可数名词，侧重体力或脑力劳动）5，7，12，20，22，24，25，27，28，29，34，44，51，55
    workplace 工作场所 20
    World Wide Web 万维网 21

译后记

近年来，金融素养已成为培养孩子全面发展的一个重要方面。早在20世纪30年代，美国就开始了对中小学生进行与生活密切相关的理财教育。如今，美国中小学理财教育日趋成熟，主要围绕让中小学生正确地"认识钱、花钱、挣钱、借钱、分享钱以及让钱增值"而展开。在英国，随着金融理财教育的需求不断上升，金融监管局将个人理财知识纳入2008年实施的《国民教育教学大纲（修订）》中，要求中小学校必须对毕业生进行良好的金融知识教育。我国周边的国家如孟加拉、斯里兰卡等，也早已开设了此类课程。

中国的孩子也同样对生活中的金融知识充满渴求。2014年春节期间，《新京报》记者调查了北京90名10～13岁的孩子，结果发现，孩子们平均收到了4 867元压岁钱，比前一年上涨了5%，其中收得最多的孩子，压岁钱有2万元，而一半以上的孩子收到的压岁钱在1 000～5 000元之间。孩子们的压岁钱该怎么处理？一部分家长的做法是直接"据为己有"：要么存入自己的银行账户，要么用到家庭的日常开支及急需的事情上。虽然也有些家长孩子的主体意识和理财意识比较强，但多局限于将孩子的压岁钱存入银行、做定投基金和购买保险等方面。其实，多数孩子都渴望由自己来管理这笔数额不少的钱，但苦于没有一定的金融和理财知识，除了交给父母或买点零食、添加一些课辅用品等之外，也不知道怎么办。因此，及时地向他们普及金融知识，让他们学会理财，应该是时候了。

华夏出版社从英国引进的"华夏少儿金融智慧屋——货币系列"丛书（共4册，中英双语）确实是应时应景之作，它涉及四个主题——世界货币、国家货币、家庭理财和个人理财，它们相互补充，构成一个整体，以孩子们喜爱的绘本形式，把晦涩难懂的国际金融、货币、贸易、经济知识转化为生动有趣的语言，用最浅显的语言全面地阐述了"金融的逻辑"，让孩子们在轻松愉悦的阅读过程中全面触摸金融知识。

完成这一系列书，我要特别感谢我的儿子贾岚晴，这套书献给已是小学生的他。我还要感谢我的先生贾拥民，感谢他一直以来对我的支持、鼓励和帮助。感谢我的母亲蒋仁娟、父亲傅美峰对我儿子的悉心照顾，使我得以安心从事翻译工作。我的朋友和同事傅晓燕、鲍玮玮、傅锐飞、傅旭飞、陈贞芳、郑文英等，也给予了我很多支持和帮助，在此一并致以诚挚的谢意！

感谢华夏出版社一直以来对我的信任！

傅瑞蓉
2015年11月于杭州

附 英文影印版

小朋友，为了方便中英文对照阅读，我们排版时尽可能使中文和英文页码一一对应，但由于中英文表达习惯不同，有个别页码的尾行可能会出现不对应的情况，这时，你只要往后翻一页就会找到哦。

小贴士

Contents

4-5 Money talk
Money is often the main topic of family conversation – after all, everyone wants a share.

6-9 Sharing it out
First you budget for the necessities. Then, if there's money to spare, the luxuries.

10-15 The BIG spend
Who makes sure the lights go on at home – and who pays?

16-21 Utilities
First you budget for the utilities that light and heat your home.

22-29 Services
Your local town or city charges taxes for the jobs it does to keep you safe, educated – and amused.

30-31 Keeping healthy
Who pays the doctor's bills? Who pays for the hospital or the ambulance?

32-33 Wash and polish
Families make a lot of mess! Who – and what – cleans up and keeps the germs at bay?

34-37 Food on the table
Food shopping is the 'BIG shop' of the week. How does the supermarket trolley get filled?

38-39 The family car
There are so many costs involved in running a car. Tax, insurance, fuel, maintenance ...!

40-41 The garden
The garden needs constant care. And maybe occasionally you'll need professional help.

42-43 Pets
Pets always seem to need a visit to the vet. And the bills aren't cheap!

44-47 Family fun
What is life without some fun? The family holiday or outing can be a major cost.

48-49 The cost of YOU
It costs a great deal of money to keep a child – yes, YOU! You'll be amazed how much!

50-51 Your education
Your education may be free or it may be paid for by your family. But there are always costs.

52-53 Your clothes
If you're a fashion freak, you'll know just how much new clothes cost.

54-55 Your pocket money
This is the important part of the family budget! Maybe you rely on this allowance.

56-57 Add it all up!
Here's a sample budget to help you tally the family expenses.

58-59 Let's discuss
Money does'nt grow on trees.
why do homes cost so much?

60-61 Glossary

62-63 Index

Money talk

How many times have you heard your dad or mum say there's just not enough money to do this or that? They just can't afford it! The subject of money and how it's spent is always around because it's money that buys everything your family needs. And this matters to you in lots of ways, right down to the family money that pays your allowance!

Why do families talk about money so much?

Very simply – it's money that decides the comfort and the style in which you live. Money gets talked about a lot because whether there's enough, or not enough, everyone in the family seems to need some.

Talking helps everyone understand what they can have and can't have – and how to be thankful when they CAN have something they want.

Money comes in

One of your parents, or maybe both, has a job. They work a minimum number of hours per week doing a special job in an office, a shop or some other place of work. They may even work at home.

The hours they work each week are paid at an agreed rate, and this is called a wage or salary. At the end of each week, or perhaps month, your parents will receive their payments. These usually go straight into the family bank account.

What jobs do your parents do?

Money goes out

This money is spent on things that the family NEEDS – the things that keep you fed, clothed and warm – and also on the things the family WANTS. The money pays for the food you buy and fuel for the car, it pays the electricity and the gas bill ... and it buys a meal if the family wants to eat out or go to the cinema, even save for a trip or holiday.

So, where does it go EXACTLY?

Sharing it out

How do your parents know what they can afford and can't afford? How do they spend the money where it's needed? Most parents set a spending pattern that decides how much money will go towards buying the family essentials each week or month. This list of really important needs, and what they cost, is called a budget.

Who's in charge?

Your parents know if they spend too much on one item they will have to spend less on another. So keeping an eye on the family budget is a key job.

You will have to ask your mum or dad who's the budget 'boss'. Maybe they do it together. Many parents share the task – and try to agree!

is there enough?

Maybe your family doesn't have a budget. Perhaps everyone just spends and hopes there will be enough to go around. Does this sound like a good plan?

It may work for a time, but then comes the unexpected. The car breaks down or the roof starts to leak. Maybe one of your parents gets ill and can't go to work. This is when the budget gets tight and everyone in the family – including you – needs to understand what's going on.

Necessity?

Necessities and luxuries

Your family's monthly expenses that help keep you clothed, warm and fed are the necessities. They are the key items in the budget.

Anything left over in the family kitty can be spent on things that are good to have, things you want but don't absolutely need, items that are known as luxuries.

Luxury?

Without a budget, it's easy to overspend. And that means getting into debt!

Debts

When you are in debt, you owe money. You may owe it to a friend or parent who doesn't care when you pay them back. But, almost certainly, when a family is in debt, it owes money to someone who DOES!

The story of Mr Micawber

A famous English writer, Charles Dickens, wrote a story about a man called Mr Micawber who spent too much.

Mr Micawber got into so much debt he was put in prison.

But he learned his lesson and had this good advice for others ...

Annual income £1
Annual expenditure 99p
Result HAPPINESS

Annual income £1
Annual expenditure £1.01
Result MISERY

Bank loans

If your family has a debt that it cannot pay, or wants to raise money for a family holiday or special purchase, your parents may talk to the bank and obtain a loan. The bank will want to know that the loan can be repaid and will set a date when this must happen. Banks also charge a borrowing fee that is called 'interest'. This is usually added to the amount of the loan and must be repaid on a regular basis.

Credit

Another way of borrowing money, where you pay much higher interest than for a bank loan, is to run up bills on a credit card. This may seem an easy way of getting extra money, but it's still a debt, and it still has to be repaid.

Is this my problem?

Debts almost always cause worry and difficulties in a family. If you understand what is going on and why, you may be able to help by tightening your belt or just being patient for a while.

The BIG spend

The home you live in may have been bought by your parents who took out a loan known as a mortgage, or it may be rented. Unless they own it outright, the cost of living in your home is almost certainly your parents' biggest expense and budget item. On average, it amounts to about 25% of the family income.

What's a mortgage?

A mortgage is a loan from a bank, or other institution, to help you purchase your home. Most banks give you 25 years or so to pay off your mortgage. The money is repaid in small amounts every month.

Banks charge their normal 'interest'. This interest fee is added to the mortgage and has to be paid back regularly as well.

What's Rent?

Some homes are rented. Rent is a fee that's paid to the actual owner, or landlord, of the property by whoever lives in it. Rent is usually paid monthly.

The rent is agreed for a period of months or years under an agreement called a lease. Once the term, or period, of the lease ends, a family can renew the lease or they must move out and find another house or apartment to live in.

insurance - what is it?

Your home, all the furniture, the appliances in the kitchen and the cars in the garage - all of these cost a lot of money. And they could all go wrong, or get broken, even be stolen - at any time. Fire, windstorms, flooding from heavy rainfall - these are all things that can damage your possessions.

Your parents will probably choose to insure the family home and everything in it. This means they pay a small sum of money each year to an insurance company which will then pay out when anything needs repairing or replacing.

upkeep or maintenance

Does the fence need a coat of paint? Must the plumber repair a leaky tap, or the builder repair a wall? These chores are necessary to keep your house and property in good order. Some of these jobs your parents can do for themselves. Others will need to be added to the family budget to pay for this repair work.

Pensions

Old and grey

Some day your parents will stop working, and once this happens, their regular wage or salary payments will stop too. However, they are probably being smart enough to save a small percentage of their pay each month. And they will go on doing this for years and years.

This money is taken from their wages or salary each month and sent to the government to hold on their behalf. Your parents may also save for their old age with other companies too.

So - what's a pension?

These savings are called a pension and will be paid back to your parents after retirement. The pensions are invested and earn money so that the total amount will grow steadily each year. They will finally add up to a sum that is larger than the total taken out of their wages.

THE WAY INTEREST WORKS

If you save your money in the bank or lend it to someone in business, you might like to see it grow. You would like to get a little more back when you are finally repaid.

One way to make sure this happens is to charge a fee known as interest. Interest is the profit or reward paid to the lender. Without interest, lenders wouldn't have any incentive to invest, and there wouldn't be much money growth.

There are different ways of adding interest but the most simple is called just that – simple interest. This just keeps adding interest at a steady rate to the original sum.

Year	Amount invested	Interest added each year 10%	How your £ grows
1	£1...............	....	....
2			£1.10
3			£1.20
4			£1.30
5			£1.40
			£1.50

 Tax

It's a sure thing that you have heard your parents groan about paying income tax. This may be because it is the law and the penalties for not paying your tax contribution can be high. But even if people groan, they do understand why the payments are necessary.

Income tax is a percentage of everyone's salary that is paid to the government. Wealthy people normally pay more tax, and poorer people pay less or perhaps nothing at all. But almost everyone puts some money into the government's 'kitty'.

• • • • • • • • • • • • • • • • • • • •

Sales tax

The government may also charge a tax on the things you buy in shops. You may have noticed this when you buy something and it costs more than it says on the label. These sales taxes may go to the government or to the state but they can be as high as 20% extra on the cost of something.

H✸w is it spent?

The reason that everyone pays tax is that everyone benefits. The government uses the tax it collects to pay for services everyone in the country needs – services such as health care, defence, education and pensions.

• •

WHERE THE MONEY GOES

Not all governments spend their money in the same way but these are some of the things they spend on.

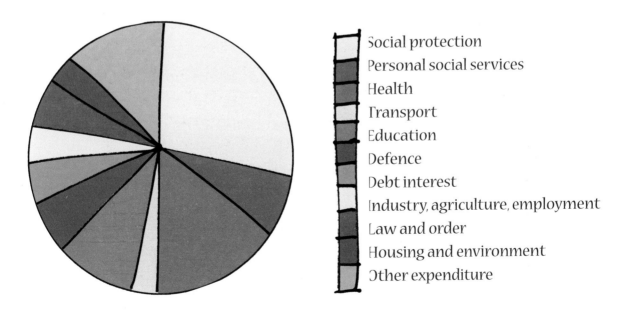

- Social protection
- Personal social services
- Health
- Transport
- Education
- Defence
- Debt interest
- Industry, agriculture, employment
- Law and order
- Housing and environment
- Other expenditure

Utilities

What are they?

'Utilities' is a useful word. It describes the basic services such as electricity, gas or water that are delivered to our homes. It also describes the companies that supply them.

Who supplies them?

It is the utility companies that make sure the lights or the TV come on, that the oven heats up, the kettle boils, or that water flows from the taps. In some countries, these essential services are controlled by, even provided by, government departments. They are so important to everyone's health and safety that the government may keep an eye on how they are run and how much they charge.

What's the cost?

The companies don't give you light and heat and water for free. They charge you.

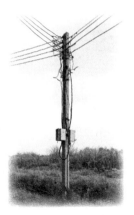

A network of poles carries power through overhead cables from town to town.

Huge power stations create electricity to send out across country to your home.

Dirty household water is carefully cleaned at the sewage plant. It is then pure enough to put back into the water system.

Oil is drilled from deep below the surface of the sea and brought ashore in pipes.

When do you pay and why?

Somewhere in your home there is a meter box for each utility. This piece of equipment is able to count how much of the service you use. It also shows how much you must pay. Every month or so, the meter will be read by a person or automatically by computer. This reading shows how much of the service you have used and how much you will then be charged. Soon after the meter is read, the household bill will arrive.

What happens if you don't pay?

The companies that supply water, gas or electricity expect to be paid for their services. They may wait if you are a little late paying, but eventually they will warn you they are switching off the supply. And they do!

Utilities: HEATING AND LIGHTING

Electricity

Electricity provides energy to light your lightbulbs and power some of your kitchen's appliances or your television. The power, which is measured in units called watts, reaches your home through cables that run from a local power generating station.

Oil

The oil has come a long way. It will have been drilled from oilfields located deep below the Earth's surface or on the seabed. Pipelines or large supertankers then carry it from country to country, sometimes half way round the world.

The electric energy will have been created by burning oil, natural gas, or coal, or it may have been created by nuclear energy or the power of falling water turning huge turbines. It is then carried great distances through overhead power lines, over fields and mountains, to your neighbourhood. In towns and cities, it travels underground or on power lines strung on tall poles, and is brought to your house.

Gas

Many homes are heated by natural gas. This flows through pipes to your home from a huge gas cylinder, or it may be brought to your household tank by truck. Like oil, it will have been drilled from under the Earth's surface and brought long distances before it reaches your home.

Water

Ever wonder where the water comes from and where it goes when you flush the toilet or wash the dishes in the sink? Water is supplied by a water company, or sometimes by your local town.

Clean drinking water comes to your house through a water pipe connected to a larger pipe buried in the street. This is known as the main and comes from a reservoir many kilometres away. This may be a lake or underground reservoir, known as an aquifer.

Dirty water leaves the house from the sink, shower or toilet and drains into a big pipe in the basement. It travels underground to an even bigger pipe under the street called a sewer.

Dirty water then flows to a sewage treatment plant which rids it of its smells and bacteria before releasing it back into the ground, river or sea.

Utilities: **KEEPING IN TOUCH**

Today, most people want to be able to talk to their family and friends, their workplace or shops and suppliers, whenever they feel like it. They also want to be able to switch on their entertainment wherever and whenever they like. Global companies supply both the machines and the connection services – and these come at a price.

Telephone

The telephone is connected to the wall by a socket called a jack. In the wall are two copper wires twisted together that reach out of the house to a phone pole. The wire travels over many poles and sometimes underground to a central switching station. From here, it fans out to other switching stations worldwide.

Mobile Phone

Mobile phone signals come to the phone from cellular towers spaced every few kilometres across the countryside. If the voice signals need to go further, they are beamed up to satellites circling the Earth and then down to a global cell tower network. Today's mobiles can do almost everything a computer can do.

internet

The Internet is a worldwide system of computers that are connected together. Computer users can communicate to billions of other users worldwide. The World Wide Web uses the structure of the Internet to enable access to linked documents, which are viewed using a web browser. Web pages may contain words, pictures, sounds and videos, and users navigate between them by clicking on links.

Television

Television signals may come through the same cable as the Internet. Some day, the screen for the computer and that for the television will be one and the same!

Letters and Parcels

The postal system in many countries goes back hundreds of years. Postmen and postwomen deliver the mail each day. However, nowadays we communicate with our phones and computers and there is less and less mail to be delivered.

EXTINCT

Services

In every village, town and city there are people who take care of certain of our needs. These are things we cannot do for ourselves, such as taking away the rubbish, providing schools and libraries, or putting out fires. These activities are called services and they are carried out by men and women who mainly work in public buildings.

RUBBISH REMOVAL & STREET CLEANING
FIRE DEPARTMENT
POLICE
ROADS
ENVIRONMENT
PUBLIC HEALTH
PARKS
LIBRARIES
PLAYSCHOOLS
THE ELDERLY
THE DISABLED
MUSEUMS
SPORTS CENTRES

A local tax

The services are paid for with money that is some kind of local tax. The tax is sometimes linked to the value of a house, or it might be based on the number of people who live there. It pays for school buildings, for teachers and for schooling. It also buys the fire engines that rush out in emergencies, the lorries that pick up the rubbish, and the salaries of everyone who works to get these jobs done.

RUBBISH REMOVAL

Every day, we toss paper and other rubbish into waste baskets throughout the house. In the kitchen, bottles, cans and food packaging are thrown into rubbish bins. All this garbage is separated into different-coloured bags for paper, plastic, metal and glass, and placed into dustbins. There's quite a stack just from one family alone.

Once a week, the garbage vanishes! The rubbish collectors come by and carry it all away. Within hours, it's on its way to be recycled or to a huge landfill site where it will be left to rot.

Rubbish must be collected before it becomes rotten or smelly and dangerous to people's health.

Thousands of tonnes of rubbish are dumped in landfill sites.

Services: **ALONG THE ROAD**

And who looks after the streets and roads? Every area has a local department called public works. This department fills in potholes, sweeps up the dirt, repaves the asphalt and, in colder climates, removes the snow with snowploughs so the traffic can flow.

In every city, clean streets mean a healthier environment.

Overhead, electricity, telephone, Internet and television lines are strung on poles. Pipes and sewers flow underground. The streets are the lifelines of a community, carrying essential services wherever they are needed. The public works department makes sure utility and other organisations keep these working.

Power and phone lines aren't always tidy but they are an essential part of home and work.

LOCAL HEALTH

Another department deals with local health. Their work is to make sure we breathe in clean air, drink clean water, and use materials to build homes, businesses and roads that are safe. This department inspects restaurant kitchens for cleanliness, and monitors the air quality for pollution.

THE COUNCIL PURSE

Money must also be found for ...

Playschools and schools
From an early age, you will be educated along with other children in your neighbourhood.

Libraries
Thousands of books and other media can be borrowed or read on a comfy seat at the library.

Parks
Open spaces with trees and play areas are laid out for everyone to enjoy.

Caring for the disabled
Disabled people have special needs at school, work and home, and the council helps.

Caring for old people
Elderly people cannot always take care of themselves, so care homes help out.

Sports centres
Join a team or exercise on your own at the sports centre.

Museums
See your favourite animals, dinosaurs, history or art in the local museum or gallery.

Services: THE FIRE DEPARTMENT

Everyone is fascinated by fire engines. It's an exciting moment when the fire alarm bell or siren starts to sound, and everyone moves to the side of the road to let the engine through.

The fire engine is called out to house fires as well as fires in factories, offices and shops. It carries long extension ladders to reach high roofs and high-pressure water hoses to quickly put out fires. These hoses connect to fire hydrants in the street which are connected to the same water main that we get our drinking water from.

Fire engines and the fire-house ambulances also run to bushfires, automobile accidents, or someone who has fallen in their home. They may even rescue your pet cat which may be stuck on the roof!

FIRE WORK

If you want to be a firefighter, there is some tough training ahead. And you have to be smart too, and have gone right through your school education. But above all, you must be strong and fit. Imagine carrying an unconscious fire victim out of the third storey of a burning building and down a ladder to safety!

Water is drawn from the nearest fire hydrant to the hoses.

Elevated ladders can reach fires in tall buildings.

Training

There's lots of training. You must learn how to put out different kinds of fires. Some fires die out if water is used on the flames. Others need special liquids and foams. Firefighters learn to use special equipment such as fire hoses, ladders, extinguishers and fire tools.

Hazards

One of the deadliest problems caused by a fire is smoke. Firefighters learn what equipment to wear and how to crawl though tight spaces, fight fires in tall buildings and deal with hazardous materials and chemicals.

Emergency help

They may need to give emergency medical treatment on the scene. But even if this isn't necessary, firefighters must be calm and reassuring and helpful.

Several types of fire vehicles will arrive at the scene of a fire.

Services: THE POLICE

Wherever you go, you will find teams of policemen and policewomen who are there to keep order. They provide protection from thieves who rob and steal, from criminals with dangerous weapons and, indeed, from anyone who wants to hurt people or damage their property.

The police do many jobs depending on where you live. Sometimes, special traffic police hand out tickets if you park too long or park in the wrong place. And they will stop and fine you if you are driving in a dangerous way.

But the police are there to help as well. They may help you cross the street or give you advice about where to go. They are trained to work as your friend and be there to protect you.

in court

People who are accused of a crime must be judged in a court of law. If they are found guilty, they may be asked to pay a fine or sent to prison. The cost of this also comes from the different taxes that your parents pay.

POLICE WORK

Police work means helping with everyday tasks as well as standing guard.

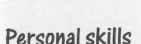

Whether on horseback or a fast motorbike, the police are watchful.

Personal skills

If you want to join the police, you will need lots of different skills. And there will be lots of exams to pass along the way! Police work can be stressful, and officers need courage and staying power!

Know it all!

The police must have basic medical knowledge to deal with emergencies, and legal knowledge to explain their actions in court. They must know how to deal with people sympathetically.

Defence

Police learn how to use firearms, defensive tactics including martial arts, how to deal with dangerous situations and people, and how not to put themselves and others in danger.

Keeping healthy

One of the most important concerns for parents is keeping the family healthy. Apart from the odd coughs and colds, and all the normal childhood diseases, your parents will make sure that any special illness is taken care of in the right way. And this kind of care can sometimes be costly.

Family Doctor

Most families register with a local medical doctor or a group of doctors known as a practice. Here, there is a file detailing all the illnesses a person suffers from birth. It also includes all the medication given – medicines, hospital care, and so on.

The local dentist has probably got the same kind of file listing all the treatment to your teeth. These kinds of records mean that the care you get is based on lots of known facts.

But all these records and visits and medicines cost money – money your family probably pays by way of special taxes.

PAYING FOR THE DOCTOR

In many countries, the government takes responsibility for making sure people are kept healthy. It requires that everyone contributes a small amount from their wages towards the cost. The service is sometimes known as national health care.

Treatment

If you contribute to a health plan you are able to see a doctor whenever you get ill. If your illness is serious, or if you have an accident, you will get free hospital treatment, even perhaps a ride in an ambulance.

Teeth

Often children will receive free dental care as well as other medical care. It is important to take care of your teeth when you are young and your dentist will tell you how to do this.

Medicines

You may need medicines to help you get better. Almost all of these will be supplied very cheaply or even free of charge under the health care plan.

Wash and polish

The one thing a family does is make a mess! The kitchen collects spilled food and grease, and dirty pans and dishes. The bathroom collects dust and hair and grease round the bath and basin. And your bedroom? Is that clean and fresh and tidy everyday?

Cleaning wages

Almost every home needs a weekly clean, which means someone has to do it. Maybe your mum or dad do the clean-up, and perhaps you lend a helping hand. Or maybe your parents employ someone who comes in and helps. If they do, these cleaning costs come out of the housekeeping budget.

Cleaning materials

There are special materials for every kind of cleaning job. There are sprays and liquids that dissolve grease, that kill germs, that freshen stale smells and help remove dirt from clothes. And lots of soaps and shampoos that clean YOU!

Gadgets and machines

The cleaning equipment in a home is a big expense. Just think how many appliances your family has that clean and iron clothes, that clean floors, walls and even clean other gadgets!

Insurance

And appliances go wrong! Perhaps the machinery inside breaks down or some small piece of the machine snaps off. Your parents may have bought maintenance insurance which will help pay for the repair, or even replace the whole gadget.

Insurance describes a way of paying regular small sums of money to an organisation called an insurance company, in case something goes wrong. The insurance takes responsibility for the costs of any breakdown if and when that happens.

Food on the table

One of the biggest costs for a family is the weekly food shop. Buying basic food items is a necessity, but of course there are 'luxury' foods as well. That's stuff you might like but don't really need, such as ice-cream or cake.

GROW YOUR OWN

Many years ago, when many folk worked on the land, people grew their own food. Today, you can still grow food in your garden. You might grow fruit, vegetables or even some grain. You might have chickens or goats or even a cow for milk. The food you produce will be cheaper than the food you have to buy.

Fruit is fresh and cheap on open air market stalls.

But there are still costs that have to be taken into consideration. You have to invest in the seeds – and the animals – in the first place. You need tools and animal feed. Even so, growing your own food is less costly than buying it.

At the Store

Although there are still open-air markets where farmers and growers sell produce they have grown themselves, most people today buy from corner shops, specialist food outlets or giant supermarkets.

Fresh carrots from the garden are packed with vitamins.

The Cost of Brands

Your parents will check the prices of each kind of food in the supermarket aisle. They will have to decide what 'brand' to buy – the name or logo given by the manufacturer – and prices vary from brand to brand. If you recognise the name of the brand, it's usually a nationally advertised one and will cost more. The store's own brand will cost less because it's not supported by television, radio or magazine advertising.

How Much?

Then there's the size of the container. A two-litre container of milk will cost less per litre than a one-litre container. But if the larger container isn't used up by a certain date it may spoil – so then it's not a bargain!

Is it Fresh?

Almost all food products are marked with a date. This shows how fresh the food is and advises the best time to buy it and the expiry date before which it should be used.

Eating Out

Mum says she can't be bothered to cook and Dad can't either, so the family will eat out. Eating out is the most expensive way to buy food. Although in some countries it's less expensive to eat at restaurants, diners, delis or fast-food joints, most families budget for maybe one or two meals out a month.

See the difference!

Made-it-at-home hamburger includes: ketchup, lettuce, tomato, bun, cheese, large hamburger
PRICE £1.03

Fast-food restaurant – same hamburger
PRICE £3.30

UPS AND DOWNS

So what we do with our money is all about choice. Or is it?

Unfortunately, spending some of our money on shelter and food is a necessity, and here we don't have many choices. In fact, most people find that most of their money is spent on necessities, so working out how much these will cost is essential.

Prices can go up and down and people's earnings can go up and down, so how do they do it? One way people keep tabs on what money can buy at any time is by using a measure called 'the cost of living'. This means just what it says – how much it costs us to live.

It's always good to check prices before you buy anything.

A basket of foods

For food, they compare the cost of a standard basket of foodstuffs and see how the cost changes. It can change from year to year, month to month or even week to week, and that makes budgeting tricky.

Even the price of processed foods can vary, although producers can pay for raw foods when prices are low but not pick them up until much later on when they need them.

All because of the weather

The price of raw foods such as grain and coffee (usually called commodities) often depends on the weather. When there is a drought in the fields where wheat is grown, the harvest yields less wheat. There is less to sell so the price goes up. This, in turn, hikes up the price of bread and cereals.

Even the price of meat is affected. That's because with less grain to feed them, cows and other farmed animals are slaughtered earlier in their lives. There are fewer beef cattle available the following year, so the price of beef goes up along with the price of your steak and hamburger.
And it's all because of a change in the weather.

Crops may be bought when young ...

... before prices rise at harvest time.

Some crops are dried and preserved.

In a severe drought, crops fail.

The family car

You probably often ask one of your parents to take you somewhere in the car. And maybe you get annoyed when they say they won't. Why should you take a bus if there's a car to be driven in? Well, the answer is that running a car is an expensive business. Putting fuel in the car is only part of the cost of running it.

Buying it
The first and probably highest cost is the initial outlay used to buy it or to lease it. Leasing is a kind of rental system. Most people cannot afford to buy a new car, or even a used car, outright. That means they have to borrow most of the money and pay it off in monthly amounts over a set period of time.

Licence
To drive a car you need a licence. This will cost a fee. Licences can last from two years to ten years, then they have to be renewed.

Registration
Each car has to be registered with a special office to prove ownership. This can cost more than the cost of a licence in some countries.

Taxes
As well as tax on fuel, the local or national government also puts a sales tax on the price of the car. This helps pay for the upkeep of roads, bridges and highways.

Fuel

Most cars today are powered by petrol or diesel fuel. The cost of that fuel depends on the price of a barrel of oil to the fuel companies and then the tax that governments put on it. In some countries, the tax is four times the price of the fuel. So smaller or hybrid-engined cars that run on a combination of petrol and battery power, are more economical than big petrol guzzlers.

Maintenance

Even new cars need to be properly looked after. And this means maintenance costs. Oil and filter changes, brake parts and tyres all have to be paid for. In some countries, after a number of years, a car has to undergo a safety check. Any part that's worn and could prove dangerous must be replaced. Replacement parts, as well as the cost of the check, have to be budgeted for.

A car dump may provide metal for recycling.

Pollution

Cars are one of the most severe causes of pollution on our planet. They are powered by fuels that give off emissions. They give off fumes that damage our health and our planet. The cars end up broken, damaged, or just plain old – and turn into huge mountains of scrap metal. Do what you can to use public transport – and your legs.

The garden

If your house has a garden, then growing some of the food you eat is a winning way of saving money – AND helping the planet. You'll help reduce the use of fossil fuels and the resulting pollution that comes from the transport of fresh produce from all over the world – in planes and refrigerated lorries – to your supermarket.

All you need is a few square metres of soil, access to water and a little time.

On the window sill

Even if you don't have a big garden – or any garden for that matter – you can still grow food. Consider container gardening if you have a sunny balcony or patio, or an indoor herb garden on a windowsill. You'll be amazed at how many tomatoes or peppers can grow out of one pot.

Improve your health

Eating more fresh fruits and vegetables is one of the most important things you can do to stay healthy. Fruit and vegetables you've grown yourself look and taste better, and their vitamin content will be at their highest levels as you bite into them straight from the garden.

Save money on groceries

The family food bill will shrink as you use produce from your backyard. A packet of seeds costs very little, can produce kilos of food – and you can save seeds from dead flowers and from inside the fruit and vegetables. You then dry them and plant them the following year.

Enjoy better-tasting food

Fresh food is the best food! How long has the food on your supermarket shelf been there? How long did it travel from the farm to your table? Comparing the flavour of a homegrown tomato with the taste of a shop-bought one is like comparing apples to wallpaper paste. Many people are concerned about food safety in our global food marketplace. When you grow your own food responsibly, you can trust that your food is safe and healthy to eat.

Pets

Your family will almost certainly have a pet animal. Over half of all American families own a pet, and over 70 million homes throughout Europe have pets.

Now, of course your pet may be as small as a gerbil – in which case, it will cost very little. But if it's a rather large hound, or even a horse, then the cost is going to appear equally large in the family budget.

PETS IN EUROPE

There are nearly a quarter of a billion pets kept in households throughout Europe. Can you find out how your country compares with the cost figures for pets in the UK on page 43?

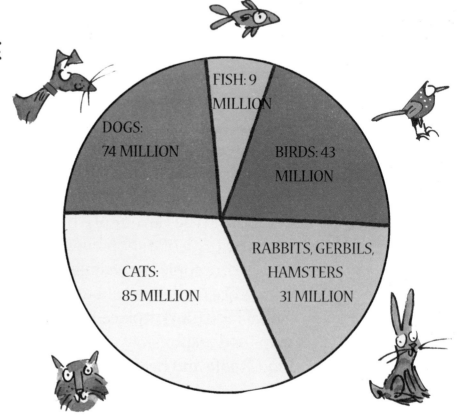

FISH: 9 MILLION
DOGS: 74 MILLION
BIRDS: 43 MILLION
RABBITS, GERBILS, HAMSTERS 31 MILLION
CATS: 85 MILLION

Breakdown of pet costs:

In addition to the initial cost of adoption – and some animals cost a lot to buy – there's a whole list of expenses that the average pet owner will incur within a year.

The estimated cost of a dog for:
Toys/Presents/Treats
Grooming
Vet fees/Medical treatment
Kennels/Catteries
Food

Total spend: £1,183

A horse can cost as much as £4000 a year

PET COUNT

It is estimated that 48%, or 13 million households, have at least one pet. There are around 67 million pets in the UK.

Dogs: 8 million
Cats: 8 million
Indoor Fish: 20 - 25 million kept in tanks
Outdoor Fish: 20 - 25 million kept in ponds

Rabbits: 1 million
Caged Birds: 1 million
Guinea Pigs: 1 million
Hamsters: Over half a million
Domestic Fowl: Over half a million
Lizards: 300,000
Frogs and toads: 200,000
Newts / Salamanders: 100,000

Snakes: 200,000
Tortoises / Turtles: 200,000
Gerbils: 100,000
Horses / Ponies as pets: 100,000
Pigeons as pets: 100,000
Insects: Less than 100,000
Mice: Less than 100,000

Family fun

Your parents work hard at their jobs. You work hard at school. And at home, there are chores to do, homework to complete, and an endless list of jobs that nobody wants to do! So free time becomes valuable.

Playing sport is a great way to spend your free time

Of course, you may want to spend it glued to your computer or lost to the world with your headphones on, but most families try to share some activities together. And lots of these are too good to miss! Trips to adventure parks, to the cinema or open-air festivals, to restaurants, racing events, football matches, and so on – why would you want to miss out?

Fun for free

Now, lots of fun is free, but lots of it isn't. In many cases, the amount of fun you can enjoy depends on how much holiday or free time your parents have, as well as how much money is available after paying the household expenses.

Fun for very little

A family can enjoy their leisure time without having to spend an awful lot of money. Playing games such as football in the back garden or the park just involve getting some friends together. And a bike ride into the country or through city streets can bring everyone together.

Free time can be spent with friends or family ...

Inexpensive fun

Here are some more ways you can enjoy your leisure time without having to spend a fortune:

Have a party and invite all your friends to bring some food and drink with them.

Spend a day at the beach or park. Have lunch at a cafe or bar.

Go to a football match or concert and sit in the cheap seats higher up. You may be far back but big screens will help you see the action.

... or on your own!

HOLIDAY TIME

Holiday time means fun for the whole family. Vacations can be taken in winter, spring or summer, and usually last a couple of weeks. You can take them at a resort, a beach house, a mountain lodge or even a cruise ship. How long you stay and how comfortable your place will be depends on what the family can afford.

Family holidays at theme parks have to be 'once-in-a-lifetime' experiences!

If your destination means travelling to another country, then costs will rise. Most families know when they'll take a holiday well in advance. So they can budget for the cost of it by putting some money away each month. You can do the same by building your own holiday fund from pocket money or money you earn.

Budget ...

When companies advertise budget holidays or budget airfares, they're talking about holidays or fares that 'suit your budget'. They're assuming your budget is limited. What they actually mean by budget is inexpensive or cheap.

Budget holidays often mean staying in your own country - maybe hiring a caravan and travelling around. It can also mean taking a package holiday in another country. On a package holiday, much of the stay will be organised for you, including your flight, hotel and meals.

... or Luxury

Then there are luxury holidays. These will almost certainly take you to some popular resort where hotels and facilities are posh. And where beach huts, entertainment and sports activities are extra.

In the end, your holiday will depend on how much your parents are willing to spend and that means setting out a budget to make sure the money is available ...

... once that's done any holiday should be fun.

The cost of YOU

It's certain your parents do not see you as a COST! They love you and want to give you the very best chance to be happy. They want you to be healthy and strong and have the best opportunities to learn and develop your skills. And they probably don't add up the money it takes to give you all these things. But add up it does!

Let's imagine your parents will look after you from the day you are born for 21 years – right through to the end of your university education.

They won't spoil you but they will cover the basic costs or necessities – like food and clothing, school uniform and daily travel. And they will pay for some luxuries too – after-school sports and clubs, a computer, a mobile, pocket money and birthday presents.

Then, in the UK, your parents will be shelving out a whopping £222,500. That's nearly a quarter of a million pounds! And in the USA, this amount would be well over $300,000.

The Cost of You

Education:
> including uniforms,
> after-school clubs and
> university costs, transport

Childcare and babysitting
Food
Clothing
Holidays
Hobbies & toys
Leisure
Pocket money
Furniture
Personal
Other
Total in $US – 9,000

Paying for childcare is a big cost for parents who work.

Developing your creative skills is an important part of your education.

Half the Children

Almost half the world – over three billion people – live on less than $2.50 a day. In fact, half of all the children in the world live in poverty.

Using $2.50 a day, the cost is roughly US$900 for raising a child for a year, and US$16,500 for raising a child from birth to age 17. Half of all children in the world live in poverty.

YOUR EDUCATION

It seems to go on forever. You get up, you go to school, you come home, you do homework. It's all about school. And sometimes you wish it would just end. But luckily, your parents have explained to you that your education is probably the most important thing in your early life.

Education for 'Free'

Fortunately for parents, early education in most countries is usually free. That includes junior or elementary school and even secondary school. The money to build schools and hire teachers comes from the taxes your parents pay either to local councils or to the national government. Taxes may also pay for other things such as bus transport, schoolbooks and sports activities.

Education Costs

Your parents know that without education your job opportunities become limited so your ability to earn money in the future is tougher. So what they spend will always seem worth it, and when you ask 'Do I HAVE to do my homework?' – you already know the answer!

School after school

In Korea, education is supported by special tutoring. Ninety-five percent of middle school pupils attend after-school tutoring agencies called 'hagwon'. These agencies help children prepare for entry into schools with high academic standards. Korean students also attend specialist academies that teach martial arts or music. So, many pupils work long hours and don't get home until late.

College and university

Where you live often determines how much you'll have to pay for a college or university education. In most countries, taxes pay for only a small portion of the cost. Parents have to pay out for most of it, or students can get a cheap student loan which must be paid off in time. In the United Kingdom, there is a limit to how much universities can charge. But it's still thousands of pounds. In the States, it can be up to four times as much, while in Indonesia, it's more affordable – just ten percent of the US cost.

If you live away from home, this adds a further cost. You have to pay for your room and your food.

One way of lowering the cost of higher education is to win a scholarship. This may provide you with extra funds for tuition and living. It may come from the government or be paid by private donation to the university.

YOUR CLOTHES

Are you a skimpy wardrobe person – or are your cupboards and drawers bulging with clothes? Or somewhere in between? You may be one of those people who think it's essential to have the most fashionable, and probably most expensive, trainers on the street. But the only really essential thing about clothes is to have some at all.

in uniform

In many countries, you will spend the greater part of your time in school uniform. This is one way to cut down on clothing expenses for young people. School uniforms seldom look fashionable but they do ensure that there's no competition to see who can dress the best and no pressure on parents to buy lots of outfits. School uniform is the great 'equaliser'.

TRENDY GOODS

Everyone knows you don't need the trendiest designer clothes to look and feel good. But this needn't mean dressing in rags. You can get the look you want without paying top prices.

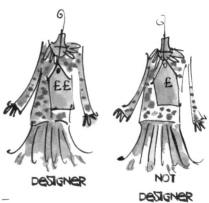

Advertisers and fashion magazines may suggest that you do, but that's their job. What you have to take into consideration is what you – or your parents – can afford. The clothing portion of your parents' budget can be around 10% of the family income – that's a lot of money.

Getting Cheaper

Clothing doesn't have to be expensive. There are designer outlets that sell at discounted prices, as well as chain stores that sell inexpensive stuff.

Shop at the stores where your money buys the most at the best value.

In general, clothing today is less expensive because so much is manufactured in countries such as China and India where labour costs are low. A T-shirt sewn in these countries can cost just half as much as one sewn in the USA.

53

YOUR POCKET MONEY

Your parents probably don't charge you to live and eat in the family home, and they may buy some of your school supplies, but most prefer to give you an allowance and let you look after your own personal expenses.

Many parents believe it's important for you to start handling your own money at a young age. They may decide to give you a regular allowance as pocket money rather than just hand out cash whenever you ask for it.

How much pocket money you get depends on how much money your parents have available to give you or how much they think you should have.

Agree on the amount

Pocket money is your first step to receiving a regular income. You may like to agree on a weekly allowance depending on how the family budget is arranged. It will also depend on how far you can trust yourself to stick to a budget. It's no good taking a weekly sum if you can't trust yourself not to blow it all in the first day!

It's possible that your pocket money comes with strings attached. These may require you to carry out certain chores around the house. You need to know EXACTLY what your parents expect of you.

Your 'Earnings'

Chores may be simply helping around the kitchen or might involve just keeping your own room clean – which you should be doing anyway! They may involve care of pets or work in the garden. They'll certainly help your parents out and involve you in taking on some of the workload of the entire household.

No problem!
You're getting paid for it!

Add it all up!

What's a budget?
A budget is an estimate of how the money that comes into the family from wages that your parents (and older brother and sisters if you have any) is to be spent. This is the family income.

What goes out over the same period of time is known as the expenses. Here's an example of a family budget.

Let's say that your family's wages come to £4,500 per month. As you can see from the chart, the budgeted household expenses come to £3,945.

The £555 that's left over can be saved or used for 'luxury' items. This might be put towards a night out at the cinema, or something your family would like to buy together – or even a small extra allowance for YOU!

INCOME per month

FAMILY WAGES £4,500

EXPENSES per month

Home costs:
Food	£ 450
Electricity	£ 200
Gas	£ 150
Water	£ 35
Repairs	£ 100
Maintenance	£ 175
Insurance	£ 180
TOTAL	£1,290

Mortgage or Rent £1,500

Travel
Auto Fuel	£ 105
Auto Main.	£ 150
Insurance	£ 125
Auto Loan	£ 275
TOTAL	£ 655

Taxes
Water	£ 50
Local taxes	£ 450
TOTAL	£ 500

GRAND TOTAL £3,945
NET AMOUNT FOR SAVINGS £555

DOWNLOAD

You don't have to write everything down on paper like this. If your parents haven't done this already, set them up with a budget template.

There are some very easy-to-use spreadsheets that can be downloaded from the web. Try:
http://www.budgetworksheets.org/

Now get your parents to help you fill in all the details and the sums will all be done for you.

Let's discuss

Does money grow on trees?
Almost everybody wishes it did. Life might be a lot simpler. And we would have a lot more forests on the planet. Unfortunately, money has a value and is exchanged for goods and services that also have a value. And it's precious stuff! Every penny you spend has to be earned – usually by your parents for each hour they work at their job.

Smaller bills?
There are many ways you can help to reduce costs in the family budget. Think about what costs there are around the house and during a typical day – electricity, gas, water, heating, travel expenses, food shopping, clothes etc. What can you do to reduce the bills? A few starter clues – switch off the lights when you don't need them, walk instead of taking the bus. Discuss other things you can do.

How do you earn money?
If you're lucky, you'll be given an allowance to spend as you like. However, many parents believe that this pocket money should be earned. You can do this by helping around the house or running errands. What jobs could you do which would help your family? You can also ask neighbours or family friends if they will pay you to do some chores.

What's wrong with shopping?
Nothing at all if you really need to buy something. But DON'T shop for entertainment! Remember: the family budget is divided into necessities and luxuries so you need to spend sensibly. The best way is to write a list and stick to it. Think of some other ways you can control your spending.

Why do you need banks?

A bank account is essential for your parents so they can put their money somewhere safe and also pay the bills. You might have a piggy bank where you keep your allowance and any spare coins – but you can also save money in the bank. If you put a little money in a savings account each week or month, the bank rewards you by paying you interest. Discuss how you can open a bank account and what you'd like to buy with your savings.

Why not create a wish list with the price of each item so you know how much you need to save in your account? You could even give yourself a deadline to earn and save the money before a certain date.

Keeping things safe?

When everything costs so much, it's really important that you look after it all. Can you remember how you can protect yourself, your money, the family property and your environment? A few starter clues – there are special ways to cover costs in case things go wrong, or if you're unwell.

Also, your environment needs to be looked after to keep costs down and to make it a healthy place to be. Discuss some ways you can protect and improve your environment. A few starter clues – you can recycle as much as possible, you can grow your own vegetables. What other ways can you think of to protect you and your family?

Credit or loans?

Your parents might also go to the bank to get a special kind of card so they can pay for things without using cash. These are called credit or debit cards. Do you know the difference between them? Discuss which is better to use and why! Lastly, banks also give loans for things that your family need, but cannot afford to buy outright – like a car. Can you remember what other loans your family might need?

Glossary

allowance
- see pocket money.

appliances
Mechanical gadgets, such as kettles and washing machines, that perform household tasks.

aquifer
A type of underground water reservoir.

bank account
A person's agreement with a bank to look after money.

bill
A statement issued by a supplier to a customer requesting payment for work or supplies.

brand
A mark or name on a manufacturer's products which is easily recognised.

budget
An agreed sum to be spent.

commodities
The name given to essential products such as grains and metals that are bought and sold in large quantities.

cost of living
The price of basic home expenses which can be compared from time to time.

credit
Money that is borrowed.

credit card
A card that lets you use borrowed money to buy goods.

debt
Borrowed money to be repaid.

disabled
This describes a person who has limited use of some part of their body and who needs special help.

essentials
Basic costs in a household. Also known as necessities.

expiry date
The date printed on certain foods before which they should be eaten.

fossil fuel
Fuels such as coal and petrol that are found in the earth.

garbage
The waste household products that are collected and destroyed or recycled.

government
The group of people elected to run a country on behalf of its people.

hagwon
A private tutoring school in Korea.

health care
A government run programme that makes sure all people receive some form of medical care.

income
Another name for earnings.

income tax
A tax paid to the government by people who earn a certain amount.

interest
A percentage sum added to borrowed or invested money.

insurance
Small sums of money paid out to cover the possible larger costs of damage or loss to property.

investment
Money lent to a person or company to help it trade.

lease
An agreement to occupy a home for a fixed period and fixed rental cost.

licence
A permission given in exchange for a payment.

luxuries
Purchases that are wanted but not needed.

meter
A gadget that counts the amount of power or water supplied to a home.

mortgage
A loan to house buyers that is repaid over time.

pension
A repayment made to elderly people by the government or a company from funds previously invested over time.

pocket money
Money paid to a child for personal expenditure.

pollution
The damage caused to the environment by human waste.

poverty
The state of not having enough money for basic needs.

price
The cost of buying something.

public works
Construction activities undertaken by a government on behalf of communities.

recycling
The use of waste to make new products.

rent
The sum paid regularly to occupy a landlord's property.

salary
Money earned on a regular basis in exchange for work.

services
Activities such as schools and museums paid for by local governments.

sewage
Waste water products from homes.

tax
A part of income or payments, collected by governments for general use.

utilities
The name given to supplies of power, water and other basic household needs.

wage
Money earned on a regular basis in exchange for work.

Index

advertising 35
afford (verb) 4, 6, 38
allowance 4, 54, 55, 56
ambulance 25, 31
appliances 11, 18, 33
aquifer 19
art gallery 25
bank account 5
bargain 35
bill 5, 9, 17, 41
brand 35
budget 6, 7, 10, 11, 32, 36, 37, 39, 42, 46, 47, 53, 55, 56, 57
cable 17, 18
car 5, 7, 38, 39
chore 11, 44, 55
cleaning 32, 33
clothes 32, 33, 48, 49, 52, 53
commodities 37
cost of living 36
credit 9
credit card 9
debt 7, 8, 9

defence 15, 29
dentist 30, 31
diesel 39
disabled 22, 25
doctor 30, 31
earnings 36, 55
education 15, 27, 48, 49, 50, 51
electricity 5, 16, 17, 18, 24, 57
essentials 7
expenses 7, 43, 44, 52, 54, 56, 57
expiry date 35
fire 11, 22, 26, 27
fire hydrant 26
food 5, 23, 32, 34, 35, 36, 37, 40, 41, 43, 48, 49, 51
fossil fuel 40
fuel 5, 38, 39, 40, 57
garbage 22, 23
garden 34, 40, 41, 45, 55
gas 5, 16, 17, 18, 19, 27, 57

government 12, 14, 15, 16, 31, 39, 50, 51
hagwon 51
health/health care 15, 16, 22, 23, 24, 30, 31, 41, 48
holiday 5, 9, 44, 46, 47, 49
hospital 30, 31
housekeeping 32
income 8, 14, 53, 55, 56
income tax 14
interest 9, 10, 12, 13
insurance 11, 33, 57
internet 21
investment 12, 13, 34
job 5, 6, 11, 22, 28, 32, 44, 50, 53
landfill 23
landlord 10
law court 28
lease 10, 38
leisure 45, 49
library 22, 25
licence 38

62

loan 9, 10, 51, 57
local tax 22
luxury/ies 5, 7, 34
mail 21
mains (water) 19, 26
market 35, 41
medicine 30, 31
meter 17
mobile phone 20
mortgage 10
museum 25
nuclear energy 18
park 25
pension 12, 13, 15
pet 42, 43, 55
petrol 39
pocket money 46, 48, 49, 54, 55
police 22, 28, 29
pollution 24, 40
poverty 49
power line 18
power station 17, 18
price 35, 36, 53
processed food 37

public works 24
recycling 23
rent 10
repairs 11, 33, 57
reservoir 19
restaurant 36, 44
salary 5, 12, 14
scholarship 5
school 22, 25, 27, 44, 48, 49, 50, 51, 52, 54
school uniform 48, 49
services 15, 16, 17, 20, 22, 24, 26, 28,
sewage 17, 19
shop 5, 14, 20, 26, 34, 35
simple interest 13
sport 22, 25, 40, 45, 47, 48, 50
supermarket 35, 40
switching station 20
tax 14, 15, 22, 28, 30, 39, 50, 51, 57
telephone 20, 21, 24

television 21
traffic police 28
turbine 18
tutoring 51
university 48, 49
utilities 16, 17, 24
vet 43
wage 5, 12
water 16, 17, 18, 19, 26, 27, 40
work 5, 7, 12, 20, 22, 24, 25, 27, 28, 29, 34, 44, 51, 55
workplace 20
World Wide Web 21

绿色印刷　保护环境　爱护健康

亲爱的读者朋友：

　　本书已入选"北京市绿色印刷工程——优秀出版物绿色印刷示范项目"。它采用绿色印刷标准印制，在封底印有"绿色印刷产品"标志。

　　按照国家环境标准（HJ2503-2011）《环境标志产品技术要求 印刷 第一部分：平版印刷》，本书选用环保型纸张、油墨、胶水等原辅材料，生产过程注重节能减排，印刷产品符合人体健康要求。

　　选择绿色印刷图书，畅享环保健康阅读！

北京市绿色印刷工程

图书在版编目（CIP）数据

家庭理财：家庭是如何花钱的？为什么要花钱？：汉、英 /（英）怀特海德著；（英）比奇插图；傅瑞蓉译 . — 北京：华夏出版社，2016.1

（华夏少儿金融智慧屋 . 货币系列）

书名原文：Family Money: How Families Spend Their Money, and Why?

ISBN 978-7-5080-8705-4

Ⅰ.①家… Ⅱ.①怀… ②比… ③傅… Ⅲ.①家庭管理—财务管理—少儿读物—汉、英 Ⅳ.① TS976.15-49

中国版本图书馆 CIP 数据核字（2015）第 306744 号

Family Money: How Families Spend Their Money, and Why?
Copyright©2014 BrambleKids Ltd
All rights reserved
The simplified Chinese translation rights arranged through Rightol Media（本书中文简体版权经由锐拓传媒取得，Email:copyright@rightol.com）
CHINESE SIMPLIFIED Language adaptation edition published by BrambleKids Ltd., and HUAXIA PUBLISHING HOUSE Copyright © 2016
All Rights Reserved

版权所有　翻版必究
北京市版权局著作权合同登记号：图字 01-2015-2438

家庭理财——家庭是如何花钱的？为什么要花钱？

作　　者	[英]威廉·怀特海德
插　　图	[英]马克·比奇
译　　者	傅瑞蓉
责任编辑	李雪飞
出版发行	华夏出版社
经　　销	新华书店
印　　装	北京中科印刷有限公司
版　　次	2016 年 1 月北京第 1 版　2016 年 1 月北京第 1 次印刷
开　　本	787×1030　1/16
印　　张	8
字　　数	140 千字
定　　价	39.80 元

华夏出版社　地址：北京市东直门外香河园北里 4 号　邮编：100028
　　　　　　　网址：www.hxph.com.cn　电话：（010）64663331（转）
若发现本版图书有印装质量问题，请与我社营销中心联系调换。